Springer Tracts in Advanced Robotics

Volume 136

The Springer Tracts in Advanced Robotics (STAR) publish new developments and advances in the fields of robotics research, rapidly and informally but with a high quality. The intent is to cover all the technical contents, applications, and multidisciplinary aspects of robotics, embedded in the fields of Mechanical Engineering, Computer Science, Electrical Engineering, Mechatronics, Control, and Life Sciences, as well as the methodologies behind them. Within the scope of the series are monographs, lecture notes, selected contributions from specialized conferences and workshops, as well as selected PhD theses.

Special offer: For all clients with a print standing order we offer free access to the electronic volumes of the Series published in the current year.

Indexed by DBLP, Compendex, EI-Compendex, SCOPUS, Zentralblatt Math, Ulrich's, MathSciNet, Current Mathematical Publications, Mathematical Reviews, MetaPress and Springerlink.

More information about this series at http://www.springer.com/series/5208

Fabrizio Caccavale · Christian Ott ·
Bernd Winkler · Zachary Taylor
Editors

Bringing Innovative Robotic Technologies from Research Labs to Industrial End-users

The Experience of the European Robotics Challenges

Springer

Editors
Fabrizio Caccavale
School of Engineering
University of Basilicata
Potenza, Italy

Christian Ott
Institute of Robotics & Mechatronics
German Aerospace Center (DLR)
Weßling, Bayern, Germany

Bernd Winkler
Department for Robot and Assistive Systems
Fraunhofer Institute for Manufacturing
Engineering and Automation IPA
Stuttgart, Baden-Württemberg, Germany

Zachary Taylor
Autonomous Systems Lab
ETH Zürich
Zürich, Switzerland

ISSN 1610-7438 ISSN 1610-742X (electronic)
Springer Tracts in Advanced Robotics
ISBN 978-3-030-34506-8 ISBN 978-3-030-34507-5 (eBook)
https://doi.org/10.1007/978-3-030-34507-5

This Springer imprint is published by the registered company Springer Nature Switzerland AG
The registered company address is: Gewerbestrasse 11, 6330 Cham, Switzerland

Series Editor's Foreword

At the dawn of the century's third decade, robotics is reaching an elevated level of maturity and continues to benefit from the advances and innovations in its enabling technologies. These all are contributing to an unprecedented effort to bringing robots to human environment in hospitals and homes, factories and schools; in the field for robots fighting fires, making goods and products, picking fruits and watering the farmland, saving time and lives. Robots today hold the promise for making a considerable impact in a wide range of real-world applications from industrial manufacturing to health care, transportation, and exploration of the deep space and sea. Tomorrow, robots will become pervasive and touch upon many aspects of modern life.

The *Springer Tracts in Advanced Robotics (STAR)* is devoted to bringing to the research community the latest advances in the robotics field on the basis of their significance and quality. Through a wide and timely dissemination of critical research developments in robotics, our objective with this series is to promote more exchanges and collaborations among the researchers in the community and contribute to further advancements in this rapidly growing field.

The volume by Fabrizio Caccavale, Christian Ott, Bernd Winkler and Zachary Taylor is an edited collection of nine authoritative contributions on the largest European programme on Robotics Challenges (EuRoC) developed through a large-scale integrating project funded by the European Commission within the 7th Framework Programme from 2014 to 2018.

The contents are divided into four parts. The first part presents the structure of the project, how EuRoC was designed and conducted, starting from the design of the Challenges, the related evaluation and selection framework to the design and implementation of the EuRoC platforms. The remaining three parts are devoted to the activities and results achieved by the Finalists in the three domains addressed by the Challenges, namely, Reconfigurable Interactive Manufacturing Cell, Shop Floor Logistics and Manipulation, and Plant Servicing and Inspection.

An effective testimony of the mission carried by the project *Bringing Innovative Robotic Technologies from Research Labs to Industrial End-users*, and thus a very fine addition to STAR!

Naples, Italy
September 2019

Bruno Siciliano
STAR Editor

Contents

Part I
The EuRoC Project: A General Perspective

Introduction

Fabrizio Caccavale

This part describes the motivations, the main objectives and the general architecture of the EuRoC project. Its aim is to provide the big picture of EuRoC as well as to share experiences and lessons learned that could be exploited for future robotics competitions/challenges.

Chapter 1 describes the design of the three challenges, starting from the main motivations that led to combine the challenge paradigm with typical features of robotic competitions. Special emphasis is put on the connection between the challenges path and the main goals of the project.

Chapter 2 is focused on the evaluation and selection activities within EuRoC, with special emphasis on aspects related to benchmarking and comparative assessment of performance.

Finally, the EuRoC platforms are described in Chap. 3. Here, the focus is on the development of environments that allow for the comparison of different methodological approaches to solving typical manufacturing use cases.

Chapter 1
The EuRoC Project: Motivations and Design of the Challenges

Anne Jan van der Meer, Fabrizio Caccavale and Gergely Eredics

Abstract This chapter describes the main motivations leading to the EuroC project and the design of the three Challenges. The design of the European Robotics Challenges (EuRoC) project was based on the belief that the definition of competitions and making them visible as "Grand Challenges" helps to create the required awareness of the core Research and Innovation issues to be solved in the community at large. In order to speed up the process of bringing innovative technologies from research labs to industrial end-users, as well as to exploit synergies across application experts, technology developers, system integrators and service providers, EuRoC launched three industry-relevant robotics Challenges within three industrially relevant scenarios. The design of the Challenges was carried out starting from the following main specifications: Adoption of an open call framework, involvement of the whole supply chain, development of each Challenge into three stages of increasing complexity. The design of the overall architecture of the Challenges was collaboratively elaborated within the EuRoC core consortium.

Keywords Manufacturing systems · Robotics competitions · Robotics challenges · End users · System integrators · Technology suppliers · Research infrastructures

---

A. J. van der Meer (✉)
CREATE Consortium, Via Claudio 21, 80125 Napoli, Italy
e-mail: jonathan@slimict.com

F. Caccavale
School of Engineering, Università degli Studi della Basilicata,
Via dell'Ateneo Lucano 10, 85100 Potenza, Italy
e-mail: fabrizio.caccavale@unibas.it

G. Eredics
InnoCentive EMEA Ltd., 344-354 Grays Inn Road, London WC1X 8BP, London, UK
e-mail: geredics@innocentive.com

F. Caccavale et al. (eds.), *Bringing Innovative Robotic Technologies from Research Labs to Industrial End-users*, Springer Tracts in Advanced Robotics 136,
https://doi.org/10.1007/978-3-030-34507-5_1

1.1 Introduction

European manufacturing industry needs competitive solutions to keep global leadership in products and services. The European society will greatly benefit from this leadership position by making use of these robotics products and services that are aiming at improving their quality of life and work conditions. The adoption of robotics technology will help European manufacturing enterprises, in particular Small and Medium Enterprises (SMEs), to adapt to global competitive pressures. ICT suppliers such as robot manufacturers, system integrators and technology solution providers are all important facilitators to improve the efficiency, adaptability and sustainability of manufacturing systems as well as their better integration within business processes in an increasingly globalised industrial context.

To boost robotics and manufacturing in Europe there is a need for a European perspective on the topics that need to be researched and developed. Both the Strategic Research Agenda (SRA) for robotics in Europe, published in July 2009 under the auspice of the European Technology Platform (EUROP), and the Factory of the Future multi-annual roadmap deliver this European perspective and are at the same time perfect examples of a successful collaboration among all the actors of the value chain in manufacturing and servicing, i.e. end users, system integrators, technology suppliers and research experts. During the development of the strategic research agendas of robotics and manufacturing, a collaborative cross-pollination across these groups led to a technology roadmap targeting at realising new product visions and applications scenarios by 2020. However, the implementation of these strategic research agendas and the realisation of the application scenarios and associated product visions also require an awareness of the core Research and Innovation issues to be solved and a joint effort of academia and industry, including SMEs.

The design of the European Robotics Challenges (EuRoC) project, funded by EU under the Seventh Framework Programme (FP7), was based on the belief that the definition of competitions and making them visible as "Grand Challenges" helps to create the required awareness in the community at large. Indeed, it is known from past experience that game-changing, competitive solutions are in the longer term very often created by the close collaboration of industry and academia. Furthermore, resorting to experimentation on shared research infrastructures, while referring to jointly developed benchmarks, is a key factor to boost the development of new solutions.

In order to speed up the process of bringing innovative technologies from research labs to industrial end-users, as well as to exploit synergies across application experts, technology developers, system integrators and service providers, EuRoC launched three industry-relevant robotics Challenges within the following scenarios:

- Reconfigurable Interactive Manifacturing Cell
- Shop floor Logistics and Manipulations
- Plant Servicing and Inspection.

The purposes of running the three Challenges were (1) to question the status quo of the existing solutions in the respective fields, (2) to showcase the potential of

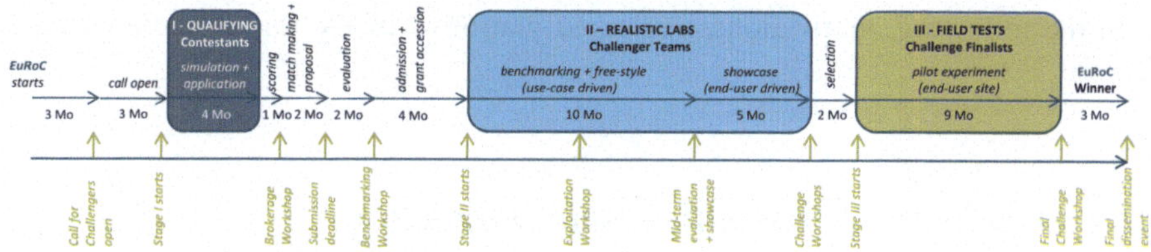

Fig. 1.1 Timeline of the challenges

robotics Challenges and bring Europe to the forefront of grant induced innovation, and (3) to develop innovative products and services that are applicable to end users and benefit the European manufacturing industry, turning innovative technologies into sustainable end user applications.

The design of the Challenges was carried out starting from the following main specifications:

- Adoption of an open call framework.
- Involvement of the whole supply chain, in order to foster collaboration among application experts, technology developers, system integrators, service providers and potential end users.
- Development of each Challenge into three stages of increasing complexity, ranging from simulation contests to real application experiments.

The timeline of the Challenges was fixed in advance and is reported in Fig. 1.1.

The design of the overall architecture of the Challenges was collaboratively elaborated within the EuRoC core consortium mainly during the Challenge Design Workshop (23–24 January 2014, Ismaning, Germany), while other meetings (e.g., the Pre-benchmarking Workshop, 4–5 May 2015, Paris, France) were devoted to the definition of details of the various stages. The main outcomes of the design process are discussed in the following.

1.2 Challenge Overview

For each Challenge, support on platforms, infrastructures and experiments was guaranteed by a local host team (**Challenge Host**), devoted to the particular Challenge. The main role of the Challenge Host was to provide the hardware and software required to enable Challengers to efficiently compete with their algorithms in the Challenges, within their domains of interest and expertise. Each host introduced the Challengers to the platforms, trained the Challengers in the integration of new technology on the platforms and organised experimental sessions. Moreover, the hosts supported the evaluation and selection activities by providing reports and information on the work carried out by teams during each stage (see Chap. 2).

In the following, motivations, goals and platforms of the three Challenges are detailed.

1.2.1 Challenge 1: Reconfigurable Interactive Manufacturing Cell

The Challenge Host was the **Fraunhofer–Gesellschaft zur Förderung der angewandten Forschung** with its **Fraunhofer Institute for Manufacturing Engineering and Automation (IPA)** in Stuttgart. The main goal of the Challenge was to develop the perception, control and cognition skills needed for safe close collaboration between humans and robots, as well as to enable multi-role and multi-arm cooperative robotic systems in an industrially relevant setting (Fig. 1.2).

The Challenge was hosted in an environment with multiple robot cell setups (lightweight robots, single-arm/dual-arm robots, multi-arm setups, etc.) for different production operations, as well as a mechanical workshop and an electronics lab for engineering of different tools and grippers.

The technology suppliers and the system integrators for this Challenge had not been decided in advance and were selected via an open call (see also Sect. 1.3).

1.2.2 Challenge 2: Shop Floor Logistics and Manipulation

The Challenge Host was the **Deutsches Zentrum für Luft und Raumfahrt** (**DLR**) in Munich. The main goal of this Challenge was to boost mobile manipulation with torque-controlled robots towards industrial maturity, and thus enable entirely new

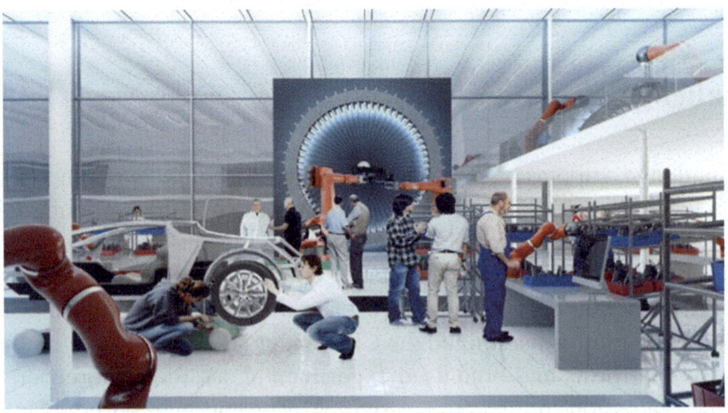

Fig. 1.2 Challenge 1: reconfigurable interactive manufacturing cell

Fig. 1.3 Challenge 2: shop floor logistics and manipulation

applications and markets, such as robots acting autonomously in unstructured environments and together with humans.

The platforms provided for this Challenge were a simulation platform, the DLR Co-Worker hardware platform and the KUKA omniRob with a KUKA LWR4+ arm, equipped with several on-board sensors.

The technology supplier and the system integrator for this Challenge were KUKA Roboter and KUKA Systems, respectively (Fig. 1.3).

1.2.3 Challenge 3: Plant Servicing and Inspection

The Challenge Host was the **Eidgenössische Technische Hochschule Zürich** (**ETHZ**) in Zurich. This challenge aimed at exploiting the great potential of Micro Aerial Vehicles (MAVs) by extending their autonomy, thus enabling their operation by untrained mission experts.

The platform for the challenge was based on the FireFly MAV from Ascending Technologies, equipped by a powerful multi-camera and other task-related sensors. The installation was set up within a flying arena, equipped with an accurate external tracking system.

The technology supplier and the system integrator for this Challenge were ALSTOM Inspection Robotics and Ascending Technologies, respectively (Fig. 1.4).

1.3 EuRoC Calls and Challenge Stages

Two EuRoC calls were issued on April 1 2014: the Call for Challengers and the Call for End Users, Technology Developers and System Integrators. 103 applications from challengers, 77 use cases, 20 technology descriptions from technology developers

Fig. 1.4 Challenge 3: plant servicing and inspection

and 12 expertise descriptions from system integrators were received. Applicant teams could be formed by research experts and solution-oriented or innovating companies (end users, technology developers, system integrators).

Each Call were complemented with a description of the Challenge Stages, the application scenarios and main tasks, the objective scoring system and the simulation environment adopted in Stage I (see Sect. 1.3.1).

Each Challenge were organized in 3 stages, described in the following.

1.3.1 Stage I: Qualifying

The Qualifying Stage was designed as a Simulation Contest, where contestants were required to solve a set of tasks by using a simulation environment defined by the EuRoC core consortium.

Admission was granted on the basis of a **1-page pitch** which included the description of the team and evidence of their skills needed to tackle the Challenge. Based on the assessment of the 1-page pitch, eligible contestants could enter the Qualifying Stage.

The designed application scenarios for this stage were:

1. **Challenge 1 (Reconfigurable Interactive Manufacturing Cell)**. The application scenario was based on the use of a dual-arm set-up, the manipulation of known objects in a structured environment and the symbiotic assembly of a product by a human worker and a robot.
    - The stage was designed to be articulated into two tracks: (1) activity recognition using pre-recorded input streams from different cameras and sensors, and from different perspectives, (2) motion planning and control of cooperative robots acting in structured environments.

- The main tasks to be solved were: Task and trajectory planning for cooperative manipulators, force/motion control for cooperative robots, object detection and recognition, human tracking, activity recognition.

2. **Challenge 2 (Shop Floor Logistics and Manipulation)**. The designed application scenario was based on the adoption of a robotic arm with 7 degrees of freedom, plus an additional linear axis mounted on a table. The set-up had to be used to perform different pick-and-place tasks, dealing with uncertain object localization, obstructed environments, as well as tasks needing higher-level execution plans.

   - The focus of the contest was on solving a series of tasks in the simulation framework set by EuRoC.
   - The main tasks to be solved were: Obtain the "robot driver's license" by running through a series of sub-tasks and solving basic problems with increasing complexity. In addition, it was required to solve various problems in the fields of path planning, object recognition and assembly execution.

3. **Challenge 3 (Plant Servicing and Inspection)**. The designed application scenario was a simplified industrial environment as well as a simulated MAV equipped with sensors (e.g., cameras, IMU). In addition, datasets with ground truth were provided.

   - The focus of the contest was on (local) map-building and handling of the MAV.
   - The main tasks to be solved were: MAV handling (state estimation, control), as well as localization and map-building, using simulated or real sensory data streams (monocular or stereo vision), and preliminary approach for generating safe paths (e.g., passing through a door or avoiding static obstacles).

The designed simulation environment for the contest was composed by 2 virtual machines (VMs). Each VM run an Ubuntu 12.04 LTS 32b operating system with ROS hydro distribution. One VM (Simulation VM) was installed with the simulation environment (Gazebo) and the scene files, along with a ROS graph to manage the simulation system and the scene. The other VM (Challenger VM) was installed with another ROS graph for control and the Gazebo viewer. It contained a set of launch scripts to start the challenge tasks. VMs were connected via internal network at startup and connected their ROS graphs. The contestants had to implement their solution on the Challenger VM. At the end of the challenge, Contestants had to submit their work by sending their Challenger VM version. This version was then run by the Challenge Host by connecting it to the Simulation VM with a modified scene file, so as to test the robustness of the solutions.

The evaluation of results and ranking of contestants were based on an objective scoring system, designed in detail and communicated in advance to contestants. The scoring system was based on the detailed definition of tasks and sub-tasks to be solved, together with benchmarks used to objectively evaluate the performance and the related metrics (quantitative criteria), scoring intervals, points, and the expected minimum level of solutions (see Fig. 1.5 for an example and Chap. 2 for further details).

TASK 1					
TASK	**SUB-TASKS**	**BENCHMARK**	**METRICS**	**SCORING INTERVALS**	**POINTS GIVEN (MAX SCORE 30 POINTS)**
Derive from human gesture where the holes for riveting are *Total 30 points*	1. Recognize pointing gesture *Total 10 points*	1.1 Success of gesture recognition (20 gesture samples)	1.1A Number of correctly recognized pointing gestures (true positives) out of 20 tests (containing 10 true positives and 10 true negatives) Penalty points apply[3]	0-4 true positives	0
				5-7 true positives	2
				8-10 true positives	5
				10 true positives	10
	2. Localize object based on CAD data Number of runs: 3, score comes from average *Total 10 points*	2.1 Accuracy of localization	2.1A Deviation in position and orientation[4]	Errors above 10 mm or 10°	0
				Errors from 6 mm to 10 mm or from 5° to 10°	2
				Errors from 3 mm to 6 mm or from 3° to 5°	5
				Errors below 3 mm or 3°	10
	3. Localize position on object at which the human is pointing Number of runs: 3, score comes from average *Total 10 points*	3.1 Accuracy of localization	3.1A Deviation in position relative to object origin	Errors above 10 mm	0
				Errors from 10 mm to 6 mm	2
				Errors from 6 mm to 3 mm	5
				Errors below 3 mm	10

Fig. 1.5 An example of task, sub-tasks, benchmarks and metrics for Stage I

The contestants were allowed to execute the simulation of tasks autonomously at their site, but they could not access the scoring software to exactly evaluate their performance (the scoring software was not provided to contestants in order to prevent optimization behaviours). They rather had to submit a log file of their simulations. The computed scores were then automatically transmitted to the Challenge Host and to the contestant; with permission of the contestant, these were also published on a public page of the EuRoC website to stimulate competition with other teams (the name of each team attached to the score was kept anonymous though).

After evaluation and ranking, **41 Contestants** were selected on the basis of the scores they obtained in the simulation tests to become prospective challengers.

As next step a list of potential end users and their use cases was provided to the prospective challengers, together with the list of interested system integrators and technology developers–those responded to the second EuRoC call. The prospective challengers were then given an opportunity to form teams with these end users, system integrators and technology developers (**match making**) at a **Brokerage Workshop** held in London Gatwick, 16–17 December 2014. With over 150 attendees this event enabled the qualified teams, end users, system integrators and technology developers to meet, discuss collaboration options and decide with whom to form candidate Challenger Teams. To facilitate the process a **playbook** of use cases, technology and expertise description they were granted access to were composed for each team.

The teams formed at this workshop then submitted short proposals with budget information (the proposal template was defined and provided by the EuRoC core consortium); in total **36 proposals** were received.

The evaluation was carried out by the Challenge Advisory Board with the help of external reviewers and independent experts. The Challenge Advisory Board was composed of 4 experts, plus an Advisor "at large" and an Advisor in representation of the European Robotics Industries.

The evaluation was based on the assessment of three criteria: scientific and technical quality of the proposal, efficiency of the implementation of the activities and potential impact of the results (see Chap. 2 for further details).

The selection of the teams who were admitted to Stage II was achieved in a panel meeting, held in Napoli (21–22 April 2015). Altogether **15** (5 for each challenge) teams were selected to become Challenger Teams. In Chap. 2, the evaluation of project proposals and selection of Challenger Teams are described in detail.

The selected Challenger Teams were contacted by the respective Challenge Hosts and asked to submit a project plan. Further to consolidation of the project plans, the challenger teams were proposed for accession to the EuRoC Grant Agreement.

Then, a **Benchmarking Workshop** (Wessling, 8–9 June 2015) was held with all the Challenger Teams. During this workshop the rules for Stage II of each challenge were presented to the teams, together with the metrics for the benchmarking round described in the following.

1.3.2 Stage II: Realistic Labs

Stage II was designed to be articulated into 2 rounds: **(a) Benchmarking + Free-style** and **(b) Showcase**. In Round (a) Challenger Teams were asked to tackle a task defined by the EuRoC core consortium (**Benchmarking round**) as well as a task largely based on their existing know–how (**Free-style round**). Round (b) concerned with an end-user driven task aimed at showcasing customizability under realistic manufacturing conditions (**Showcase round**).

Each Challenger Team received a financial support of up to €375,000, i.e. €250,000 for Round (a) and €125,000 for Round (b). The support for Round (b) was granted after assessment of the work performed in Round (a).

The designed application scenarios for this stage were:

1. **Challenge 1 (Reconfigurable Interactive Manufacturing Cell)**. The focus was on safety, reconfigurability and cost-effectiveness of the application solution. The Benchmarking tasks were designed to be focused on the safety of the worker and on measuring the economic and technical target improvements (cost, cycle times, etc.), in comparison to the performance of current solutions.
2. **Challenge 2 (Shop Floor Logistics and Manipulation)**. The focus was on typical tasks for mobile manipulators in a transformable production environment. The setup at the challenge host modeled a typical shop-floor with part storage loca-

tions, simulated production machines and assembly stations. The mobile manipulator had to perform part logistics and machine tending tasks, but it was also free to provide assistive assembly tasks up to a completely autonomous handling of the production scenario. For a successful execution of the tasks, the capabilities demonstrated in Stage I had to be used. In the application scenario, the mobile manipulator served as an advanced part logistics system, picking up parts from machines and providing those parts at assembly stations and other machines for refilling buffers. Providing capabilities for executing pre-assembly and assembly steps was encouraged for further improving the effectiveness of the overall production scenario. Human workers in the area had to be taken into account, interaction with workers could also be targeted. In sum, the Benchmarking round was focused on: recognition of parts in buffers and fixtures, delivery of said parts to appropriate assembly stations and to other machines, by taking into account the presence of human workers in the area.
3. **Challenge 3 (Plant Servicing and Inspection)**. The application scenario included collision-free navigation in a complex GPS-restricted environment and task allocation by human operator using on-board camera or direct view. The Benchmarking round was designed to be focused on tasks that are necessary to keep a MAV safe and stable, such that an operator without piloting skills would be able to perform inspection tasks within a small-to-medium-sized environment. These tasks included: control and local state estimation, local map building and local path planning, reactive collision avoidance and task allocation by human operator.

The Free-style demo had to be proposed by the Challenger Teams, while the Showcase round was to be designed according to the end-user specifications.

Each Challenge Host set a timeline and an access plan to the platforms, and organised meetings to instruct the Challenger Teams on the platforms usage and safety procedures.

The evaluation of results obtained in the Benchmarking round by the Challenger Teams was based on an objective scoring system, designed in detail and communicated in advance to the teams. As in Stage I, the scoring system was based on the detailed definition of tasks and sub-tasks to be solved, together with benchmarks used to objectively evaluate the performance and the related metrics (quantitative criteria), scoring intervals, points, and the expected minimum level of solutions (see Fig. 1.6 for an example and Chap. 2 for further details).

The evaluation of the results obtained in the Free-style round was based on two factors: the extent to which the Challengers met measurable goals/metrics determined by each team and the difficulty of the proposed tasks, measured on a scale determined in advance by an Evaluation Board, composed by the Challenge Advisory Board and additional independent experts. The evaluation of the results achieved in Round (b) was based on the ex post assessment of the same criteria used for evaluation of proposals and selection of teams admitted to Stage II, i.e., scientific and technical quality of the achieved results, efficiency of the implementation of the activities and impact of the results. A detailed description of the evaluation rules is reported in Chap. 2 for further details.

TASK 3					
TASK	**SUB-TASKS**	**BENCHMARK**	**METRICS**	**SCORING INTERVALS**	**POINTS GIVEN**
Waypoint navigation *Total points: 23* *(Points are halved if motion-capture is used)* *Weighting: 23 %*	1. Reach first waypoint within a tolerance volume on a collision-free path *Total points: 2.3* *Weighting: 10%*	1.1 Planning time and optimality of the planned path.	1.1A Time from receiving the waypoint until the vehicle arrives within a tolerance volume around the waypoint.	All other teams	Relative scoring
				Fastest team	2.3
	2. Reach second waypoint within a tolerance volume on a collision-free path *Total points: 4.6* *Weighting: 20%*	2.1 Planning time and optimality of the planned path.	2.1A Time from receiving the waypoint until the vehicle arrives within a tolerance volume around the waypoint.	All other teams	Relative scoring
				Fastest team	4.6
	3. Reach third waypoint within a tolerance volume on a collision-free path *Total points: 6.9* *Weighting: 30%*	3.1 Planning time and optimality of the planned path.	3.1A Time from receiving the waypoint until the vehicle arrives within a tolerance volume around the waypoint.	All other teams	Relative scoring
				Fastest team	6.9
	4. Reach fourth waypoint within a tolerance volume on a collision-free path *Total points: 9.2* *Weighting: 40%*	4.1 Planning time and optimality of the planned path.	4.1A Time from receiving the waypoint until the vehicle arrives within a tolerance volume around the waypoint.	All other teams	Relative scoring
				Fastest team	9.2

Fig. 1.6 An example of task, sub-tasks, benchmarks and metrics for the Benchmarking round of Stage II

A mid-term evaluation event (held in Zurich, 19–20 September 2016) after Round (a) allowed for a comparative assessment of the results obtained by the Challenger Teams and decided teams to be admitted to Round (b).

After conclusion of Round (b), Challenger Teams were asked to prepare a detailed report of the activities and the results achieved in Stage II, while the Challenge Hosts prepared a brief report describing the overall activity of each Challenger Team during the two rounds. Both reports, based on templates set by the EuRoC core Consortium, were sent to the experts acting as evaluators. A Challenge Workshop (Stuttgart, 15–17 May 2017) was held, where Challenger Teams were ranked by an evaluation panel and the 6 Challenge Finalists (2 for each Challenge) were selected for Stage III (Field Tests).

1.3.3 Stage III: Field Tests

In the Field Tests Stage, the selected Challenge Finalists were required to develop a pilot experiment in a real environment at the end-user site, aimed at achieving robust and reusable solutions to real industrial needs. The general solutions developed during Stage II had to be customised for end users and tested on the field. Thus, the field tests were carried out under real manufacturing conditions, and showcased at the end-user site.

Each team of Challenge Finalists received up to €210,000, under the obligation to prove reusability and adaptability of the benchmarked solutions in practice.

Evaluation of the results obtained by the teams during the Field Tests stage was done in conjunction with the Final Challenge Workshop, held in Napoli on 5–6 April 2018, by a Board of Judges (BoJ). The BoJ was composed of the Challenge Advisor at Large, the Challenge Advisor for European Robotics Industries, along with 5 external stakeholders: a representative from the field of policy making or government, a representative from a large manufacturing company, a representative from an SME, a representative of the entrepreneur/venture capitalist community, and a media/social expert.

Judging was based on criteria such as technical feasibility, market and competition, potential of exploitation, benefit to the society, team skills, etc. The EuRoC Winner selected by the Board of Judges was publicly announced and awarded during the Automatica Fair (Munich, 19–22 June 2018).

1.4 Conclusion

In this Chapter the main motivations and goals of the EuRoC project have been presented, and the outcomes of the design of the challenges have been briefly reviewed. It is worth remarking that, although the general structure of the Challenges and the main processes had been determined in advance (i.e., before the calls were issued), the design process continued along the whole life of the project. Indeed, lessons learned within one stage, feedback received from Challengers and/or necessary adaptions of the process often determined refinements and improvements of the design.

Chapter 2
Evaluation and Selection Activities in EuRoC: Innovations and Lessons Learned

Ramez Awad, Fabrizio Caccavale and Anne Jan van der Meer

Abstract This chapter describes the evaluation activities carried out during the three stages of EuRoC. Evaluation and selection activities in EuRoC have represented a central issue. In fact, the successful development of innovative and effective solutions of industrially relevant problems was an essential goal, from which the need to select the teams with the strongest skills and motivations, as well as the use cases with the highest possible impact. Of course, innovative approaches to compare and benchmark skills and solutions were needed. Such approaches are described in detail in the Chapter, by properly highlighting innovative solutions that can be exploited in future initiatives.

Keywords Robotics competitions · Robotics challenges · Benchmarking · Selection · Evaluation

2.1 Introduction

Evaluation and selection activities in EuRoC have represented a central issue. In fact, the main objective of EuRoC was to strengthen collaboration and cross-fertilisation between the industrial and the research community via the solution of challenging real-world problems in different industrial scenarios. Thus, the successful develop-

R. Awad
Department for Robot and Assistive Systems, Fraunhofer Institute for Manufacturing Engineering and Automation (IPA), Nobelstraße 12, 70569 Stuttgart, Germany
e-mail: Ramez.Awad@ipa.fraunhofer.de

F. Caccavale (✉)
School of Engineering, Università degli Studi della Basilicata, Via dell'Ateneo Lucano 10, 85100 Potenza, Italy
e-mail: fabrizio.caccavale@unibas.it

A. J. van der Meer
CREATE Consortium, Via Claudio 21, 80125 Napoli, Italy
e-mail: jonathan@slimict.com

F. Caccavale et al. (eds.), *Bringing Innovative Robotic Technologies from Research Labs to Industrial End-users*, Springer Tracts in Advanced Robotics 136,
https://doi.org/10.1007/978-3-030-34507-5_2

ment of innovative and effective solutions was an essential goal, from which the need to select, in the different stages, the teams with the strongest skills and motivations, as well as the use cases with the highest possible impact. Of course, selection needed comparison and benchmarking of skills and solutions.

On the other hand, EuRoC has become a "laboratory" where innovative evaluation and benchmarking procedures have been designed and could be adopted and/or extended in the framework of future initiatives.

In the following, the evaluation activities carried out during the three stages of EuRoC will be examined in detail.

2.2 Evaluation Activities in Stage I

2.2.1 Simulation Contest: Scoring and Ranking of Contestants

The criteria and the objective scoring system adopted to evaluate the results obtained by contestants in Stage I—Qualifying (Simulation contest) of EuRoC was designed by the EuRoC core consortium and then reviewed by 6 (2 per Challenge) experts external to EuRoC, to assess the fairness and objectiveness of the criteria, the metrics and the scoring system. Suggestions of the experts were integrated into the final version.

The scoring system was based on tasks and sub-tasks to be performed by Contestants, as well as on pre-defined benchmarks, metrics, scoring intervals, points, and expected minimum level of solutions. An example of scoring system is given in Fig. 2.1.

For each Challenge, the simulation software was composed of two virtualized operating systems: one containing the simulation environment, and the other the control development kit for the Challengers to implement their solution (see Sect. 1.3.1 in Chapter 1).

Based on the objective scoring system, all Contestants have been automatically ranked by the simulation software, without operator intervention. The 15 Contestants of each challenge that scored the highest number of points across all the tasks in the respective challenge advanced from the simulation contest to the matchmaking. It was decided that if less than 15 Contestants per challenge had reached the expected minimum level of solutions, the 15 best scoring Contestants still would have advanced. In case of ex-aequo, i.e. two or more Contestants had finished both at the 15th place in the ranking, to break the tie, the Contestants would have been compared on the number of sub-tasks in which they achieved the maximum obtainable scores.

An intermediate evaluation has been carried out during the contest, where contestants could submit part of the work to check if the proposed solutions worked properly. A forum and a mailing list were setup for technical support to contestants.

TASK 2					
TASK	**SUB-TASKS**	**BENCHMARK**	**METRICS**	**SCORING INTERVALS**	**POINTS GIVEN (MAX SCORE 30 POINTS)**
Carry out riveting on the box during spontaneous interference of the human *Total 30 points*	1. Fine localization of holes based on robot mounted 2D camera Number of runs: 3, score comes from average *Total 10 points*	1.1 Accuracy of localization	1.1A Deviation in position and orientation[3]	Errors above 5mm or 5°	0
				Errors from 1.5 to 5 mm or from 3° to 5°	2
				Errors from 0.3 mm to 1.5 mm or from 1° to 3°	5
				Errors below 0.3 mm or 1°	10
	2. Mount the rivets without colliding with surface *Total 10 points*	2.1 Successful mounting of rivets	2.1A Number of rivets correctly mounted in T_{max} minutes[4]	0-4 correct rivets	0
				5-9 correct rivets	2
				10-14 correct rivets	5
				15 or more correct rivets	10
	3. Avoid contact with human during robot motion *Total 10 points*	3.1 Success of human safety	3.1A Number of human collisions (during robot motion)	>0 collisions	0
				0 collisions	10

Fig. 2.1 Example of scoring system for Stage I

Challenge 1 was organized into two separate tracks: Perception and Control. The scenarios for the two tracks were simulated in ROS/Gazebo. ROS-based tools, allowing selection or termination of tasks, selection of scenarios and provision of status information, were made available to Contestants. The task were (Fig. 2.2):

- Track 1: Perception. The simulated setup consisted of a 6-DoF (Degree-of-Freedom) industrial robot mounted on a table. The robot was equipped at its end effector with a riveting tool and a stereo camera. An object was placed on the table in the working range of the robot; the object had holes to put rivets in. A 3D camera was mounted above the workplace to oversee everything on the table. A second 3D camera was mounted to capture the human approaching. The task were:
  - Task 1 required to identify the location of the holes for riveting based on a human pointing gesture. The task consisted of three sub-tasks: (1) recognizing pointing gesture, (2) localizing objects based on CAD data and (3) localizing position on object at which the human was pointing.
  - Task 2 required to carry out riveting on the object during spontaneous interference of a human. The task consisted of three sub-tasks: (1) fine localization of holes based on robot mounted 2D camera, (2) mount the rivets and (3) avoid contact with the human during robot motion.
- Track 2: Control. The simulated setup consisted of two 6-DoF robot arms, carrying objects separately (first task) as well as carrying the same object grasped tightly (second task). Both robots were equipped with joint position sensors and wrist

force/torque sensors. Poses of the objects with respect to robot TCP (Tool Center Point) frames were known with uncertainties. Dynamic parameters of robots and objects were known with uncertainties.

- In Task 3 one robot carried a peg and the other carried a hole, thus simulating a parts mating task. The aim of the task was to assemble the parts, while limiting the interaction forces during the assembly operation.
- In Task 4, the two robot arms tightly grasped a rigid object, which had to be transported from a given pose to a desired goal pose, while limiting internal stresses on it and avoiding fixed obstacles with known positions.

Of the 30 registered teams, 9 participated to the intermediate evaluation and 11 submitted their solutions to final evaluation. The team activity over time was relatively low for the first half of the contest, while a strong gain in activity was observed in the second half. The high dropout rate likely resulted from the ease of registering compared to actually solving the tasks. From several teams that dropped out feedback was received that they left for lack of time and man power.

Fig. 2.2 Simulation scenarios for Challenge 1: Track 1 (top) and Track 2 (bottom)

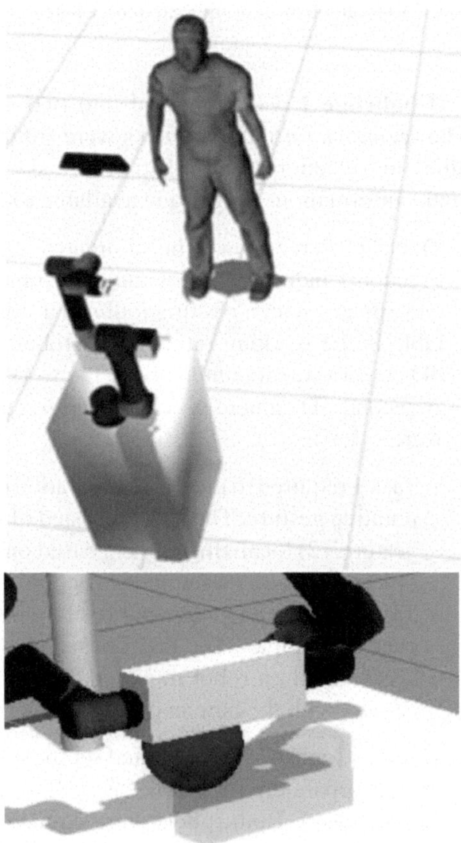

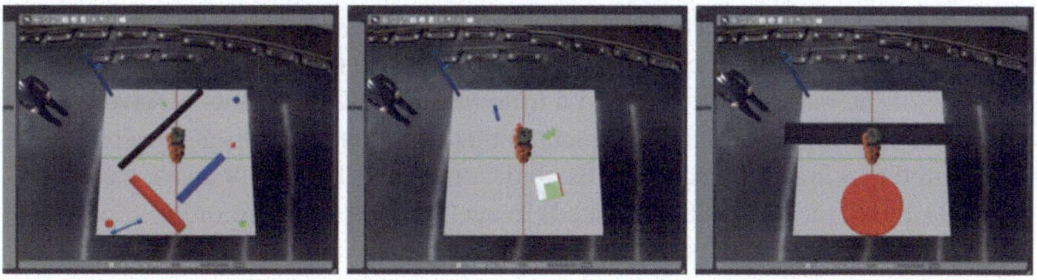

Fig. 2.3 Simulation scenario for Challenge 2: Manipulator and objects used in Task 4, 5 and 6

Fig. 2.4 Simulation scenarios for Challenge 3: Example of dataset for Track 1 (top), screenshot of the simulation scenario for Track 2 (bottom)

Challenge 2 consisted of a series of pick-and-place operations. Contestants had to pick precisely or imprecisely placed objects up from a table, and place them on the other side of the table. Each task had a time limit, and Contestants received points for correctly picking and placing objects in this time frame. Tasks 1–3 were similar and required picking and placing skills, and Contestants had to achieve a basic competence (1 block picked, 1 block placed) in each of these tasks in order to progress. Tasks 4–6 added in elements to test mobility, planning and visual tracking, respectively. The scenarios were simulated in Gazebo with a special plugin to simulate the robot. The simulated scenario consisted of a light-weight robot arm with 7 DoFs mounted on a table. The base of the robot was mounted on two additional axes which allowed the robot to move in the table plane to expand the workspace of the arm. A parallel jaw gripper was mounted at the TCP of the robot. The robot simulated the torque sensors of the joints and reacted to contacts of the gripper and/or of the arm structure with the environment, i.e., a compliance similar to that of the real robot was simulated. For capturing the environment, vision sensors (2D, 2.5D depth sensors) observing the table and a sensor at the TCP were made available to recognize and localize movable objects in the scene (on the table). For the different tasks simple basic object geometries could be inserted in the scene and manipulated by the robot. The objects were physically simulated (i.e., gravity and friction were included in the simulation). The six designed task were (Fig. 2.3):

- Task 1 required Contestants to pick and place objects (rearrange movable objects in the scene) in a precisely observable and calibrated environment.
- Task 2 followed the same format as Task 1, with the difference that there was uncertainty in object position and robot precision (the robot operating in an imprecisely observable and not calibrated environment). Contestants had to execute three runs with deviation in the placement of the objects, according to usually observable uncertainties for object localisation in real environments, and the final score was taken from the average scores of these runs.
- Task 3 followed a similar format as Task 2, but the difference that the arm base could move in the table plan along the two additional axes, since either the pick or place position was not within range of the fixed robot arm.
- In Task 4, the robot arm had mobility as in Task 3, but in an obstructed environment (e.g., a wall between the robot and the objects). Both the objects and the obstruction were imprecisely placed. Again, the arm had to pick and place the objects into the correct zones. Unintended collisions resulted in exceeded torque limits and stopped the simulation. To evaluate the generality of the obstacle avoidance strategy three runs with different obstacles had to be solved. The final score was taken from the average scores of these runs.
- Task 5 required the robot to perform loose assembly of basic geometric parts to compose a 4×4 field of basic blocks.
- In Task 6, the robot had to pick an object from a moving conveyor belt in 10 rounds with increasing speed of the conveyor belt. The same object had to be picked in each round, but its placement on the conveyor belt was imprecise and varying.

This time, grasped objects could be dropped into a basket (requiring less accuracy in the final placement than previous tasks).

Of the 39 registered teams, 17 submitted a solution to the final evaluation. The results of the final evaluation showed that the tasks were challenging but solvable. No team achieved the maximum score and almost every team achieved some points in at least one task.

Challenge 3 was organized into two separate tracks: the first track covered perception, the second track covered state estimation, control and navigation. Two testbeds were prepared for the challenge and the tasks in the tracks differed significantly. Scoring of the tracks was designed in a way that both Contestants being excellent in one track and Contestants being good in both tracks could achieve competitive scores. This aimed at bringing in experts in one field as well as all-rounders. The task were (Fig. 2.4):

- Track 1: Vision-based localization and reconstruction. In this track, rather than on a simulation environment, benchmarking was based on dataset evaluation, so as to evaluate contestants' abilities with respect to the main perception tasks that are required in order to keep a Micro Aerial Vehicle (MAV) airborne safely in a GPS (or in general an external tracking system) restricted area. The main exteroceptive sensor in this task was a stereo camera. The scenario was a real environment, in which datasets that had to be evaluated by the Contestants were collected. The simulated setup consisted of a MAV equipped with a stereo camera and an IMU (Inertial Measurement Unit) consisting of a 3 axis linear acceleration sensor and a 3 axis angular rate sensor. Typical MAV datasets with ground truth were recorded, and with different levels of difficulty. There was a first set (low to high difficulty) of datasets, where vision-based localization was evaluated, and a second set of datasets used to evaluate 3D reconstruction of the environment. IMU data could be used to improve the vision-based localization/reconstruction approaches. In both tasks, the focus was on applicability on a real system.
  - In Task 1, the goal was to localize the MAV with respect to the starting location, using camera images and IMU data only. This information was necessary for a controller to keep the MAV airborne. The dataset consisted of a set of stereo images recorded at a low rate and IMU data recorded at high rate, synchronized with the images. The focus was on speed and local accuracy rather than on global consistency. Therefore, accuracy of position with respect to ground truth in subsections of the flown path was evaluated. Since real-time is crucial for stable flight, the runtime for every single frame was evaluated and runtimes that exceeded the time between two images were penalized.
  - Task 2 was focused on detection and avoidance of obstacles. Detection of obstacles, or more in general, reconstruction of the environment was required in this task. Another set of datasets was provided, which contained stereo images and IMU data similar to Task 1. In addition, the full 6-DoF poses at which the images were captured were provided. The goal was to create an occupancy grid of the environment, which could be used for obstacle avoidance. Real-time constraints

had to be met as well, not for single frames, but within subsections of the flown path.

- Track 2: State estimation, control and navigation. The simulated setup included a quadrotor helicopter with simplified dynamics, equipped with one or two generic 6-DoF pose sensors and an IMU consisting of a 3-axis linear accelerometer and a 3-axis gyroscope; the pose sensors emulated pose measurements from a vision-based localization approach. These sensors were not aligned with the IMU frame of reference and yielded low-rate pose measurements, corrupted by noise and delay. An exact timestamp of the measurements was provided. The frame of reference of these sensors was at an arbitrary location and orientation, and measurements may not be available in the whole environment. At least one of the two sensors yielded measurements, i.e. there was no time without any pose measurement, so as to emulate a setup where the MAV is equipped with two (stereo) cameras, where it is not guaranteed that both can always localize. The control of the vehicle could be accessed at attitude control level, angular rate control level or direct motor command level. For all access levels, actuator saturations and time constants were specified.
  - The goal of Task 3 was to show basic functionality of a MAV. The vision-based localization was replaced by the generic 6-DoF pose sensor for simplicity. The noisy and delayed measurements had to be handled, and a hovering controller had to be designed (subtask 1) that could cope with constant (subtask 2) and abrupt (subtask 3) disturbances (e.g., constant side wind/wind gusts, respectively).
  - Task 4 extended Task 3 towards simple waypoint navigation first (subtask 1). In a further step (subtask 2), an additional sensor was added, since the first sensor was not able to localize everywhere along the path towards the next waypoint. Therefore, the challenge was to use information from both sensors, in order to arrive at the waypoint successfully. In the third subtask, an occupancy grid representation of the environment (similar to the one from Task 2) was provided, and the objective was to reach a goal while circumnavigating the obstacles in the map. The obstacles were simple, but placed at different heights, such that 3D planning was necessary.

Of the 34 registered teams, 18 submitted their solution to the intermediate evaluation and 21 submitted a solution to the final evaluation. A regular activity of 20 teams was observed. All tasks build on each other in real environments, but building blocks were separated/abstracted in the simulation, such that challengers were able to tackle individual tasks separately and work on one problem at a time. From feedback of the challengers this was very well perceived and appreciated.

In total, 41 teams qualified to advance to match-making phase phase, where they have the opportunity to team up with end users, system integrators and technology suppliers: 11 in Challenge 1, 15 in Challenge 2 and 15 in Challenge 3.

2.2.2 *Admission to Stage II: Evaluation of Short Proposals*

After the match-making phase, the qualified teams were required to submit a short proposal on the basis of a template provided by EuRoC. A proposal for admission to Stage II consisted of three sections: **Excellence (Scientific and technical quality)**, **Implementation** and **Impact**. The three sections were directly related to the three criteria adopted for the evaluation, set and communicated beforehand to prospective challengers:

- **Scientific and/or technological excellence** relevant to the Challenge scenario and EuRoC general objectives, novelty of the proposed application (progress beyond the state of the art), technical soundness of the concept, quality and level of difficulty of the use case.
- **Quality and efficiency of the implementation and the management**, intended to assess the efficient use of resources, quality and relevant experience of the individual participants (evaluated on the basis of the rank achieved in the simulation contest as well), quality and strength of the team as a whole, appropriateness of the allocation and justification of the resources to be committed.
- **Potential impact** through the development, dissemination and use of project results. As for this criterion, the following qualities served as indicators:
  - potential market for the developed application,
  - potential of the proposed work to impact on UE economical, technological and scientific growth,
  - potential of the proposed work to contribute to new products/services/tools in a reasonable time frame,
  - commitment of the end user to exploit the results future products/services.

To each criterion a score between 0 and 5 (use of half points was allowed) was assigned; the total score of the proposal was the sum of the three scores. A proposal could be selected only if the score for each criterion was above a threshold of 3 and the sum of the three marks was not less than 10.

The evaluation was carried out by independent experts (external to EuRoC core consortium), whose appointment was approved by the European Commission (EC); detailed guidelines for the experts acting as evaluators were provided. Moreover, evaluators were strongly advised to take into account that one of the main aims of EuRoC was to develop innovative products and services that are applicable to end users and benefit the European manufacturing industry. Thus, the activities to be carried out by teams dealt only in part with pure or applied research, while the focus was on the applicability of the developed solutions in strongly motivated use cases.

In total 36 proposals have been submitted: 10 in Challenge 1, 12 in Challenge 2 and 14 in Challenge 3.

In detail, the evaluation process was articulated into 3 steps:

1. **Initial evaluations**. For each proposal, two individual evaluation reports with scores were prepared by two independent experts (*evaluators*). This step was conducted remotely through a web platform set by EuRoC.

2. **Consensus**. A consensus report was written by a *rapporteur*, based on the individual reports. The rapporteur was in charge of writing a provisional summary of the individual evaluations (i.e., a provisional Evaluation Summary Report, ESR). The evaluators had to explicitly agree on both the consensus scores and the ESR. The ESR and the consensus marks were formed by discussion in a consensus blog held via web platform. With the exception of one proposal, consensus was reached on all ESRs and scores. The ESR and scores of the one proposal were left to the *Panel* (see next item) to decide on.
3. Ranking and selection of the proposals was achieved in a Panel meeting (Napoli, 21–22 April 2015), attended by the Challenge Advisory Board (composed by the Challenge Advisor at-Large, the three Challenge Advisors and the Challenge Advisor for European Robotics Industries), four independent experts, the Project Officer (PO) and two members internal to EuRoC (both the PO and the internal representatives acted as observers and did not take part to the evaluation). The initial ranking of proposals was based on the provisional ESRs produced in the previous step. The panel:
    - checked the consistency of the scoring (e.g., in the case two experts provided substantially different evaluations), with the possibility to increase or decrease the scores;
    - resolved cases in which proposals had equal scores;
    - reached a consensus on both ESRs and scores related to those proposals for which a consensus was not reached during the second evaluation step;
    - finalized the ESRs;
    - established the final ranking of all proposals above threshold;
    - prepared the evaluation report for the European Commission.

The ESR of each proposal was sent to the corresponding Challenge Team Leader. The ESRs and evaluation panel minutes were forwarded to the EC by EuRoC. Based on this information the EC approved the final list of selected proposals (5 for each challenge).

Of the 36 submitted proposals, 23 were ranked above thresholds: 6 for Challenge 1, 8 for Challenge 2, 9 for Challenge 3; for each Challenge, the first 5 proposals in the ranking were proposed to EC for admission to Stage II and funding.

After EC approval of the panel outcomes, evaluation and selection results were communicated to the proposers and the final ranking of the proposals was published on the EuRoC web site. The proposers were informed that a request for redress could be submitted to EuRoC. Over the 36 evaluated proposals, 3 requests for redress were received (1 for each challenge). The procedure adopted to handle the redress requests is described in the following:

- The requests were examined by a Redress Committee internal to EuRoC, composed by EuRoC team members not directly involved in the organization and management of the evaluation activities, which prepared a report with its findings and recommended a course of action. Three recommendations were possible: (a) the complaint is rejected as unfounded, (b) the complaint is upheld but the problem concerned did not jeopardise the decision whether or not to fund the proposal, (c) the complaint is upheld and a re-evaluation is recommended.

- The EuRoC Coordinator sent the Redress Committee's reports and recommendations to the EC for verification and EC opinion.
- An EC FP7 Redress Committee was convened to examine each case and confirm or reject the course of action recommended by the EuRoC Redress Committee.
- A reply was sent by EuRoC to the applicants.

The 3 redress requests were rejected by EC.

2.3 Evaluation Activities in Stage II

The evaluation activities in Stage II have been perhaps the most challenging and innovative. They were articulated into two main steps.

First, a **mid-term evaluation** took place after Round (a), so as to allow for a comparative assessment of the results obtained by Challenger Teams and decide admission to Round (b).

Then, evaluation of the results obtained by teams in Round (b) and admission to Stage III was decided in a **Challenge Workshop**.

Independent experts (external to EuRoC) were selected for the Evaluation Board, in charge of evaluating the results obtained by the Challengers during Stage II and deciding admission to Stage III. The Evaluation Board was completely independent from the EuRoC core consortium and was composed by the Challenge Advisory Board and additional independent experts.

2.3.1 Benchmarking Round

The evaluation was based on the results obtained by Challengers, measured through objective metrics and a scoring system similar to that adopted in Stage I. The evaluation runs were carried out on the three EuRoC platforms (see Sect. 1.1.3 for technical details on the platforms), according to a schedule determined by the Challenge Hosts.

For evaluation purposes, the Challenge Hosts captured official video footage and relevant data during the evaluation runs. The Challengers were allowed to take their own videos and edit them, but this material was not considered for evaluation purposes, but only for dissemination purposes. Moreover, Challenge Hosts prepared a report describing the overall activity and the results of the Challenger Team during the Benchmarking round. These reports were made available to the Evaluation Board (together with official video footage and data captured during the evaluation runs).

In **Challenge 1**, the scenario was based of a car door and module platform (Fig. 2.5). The robot platform was composed by a UR10 robot manipulator, 3D sensors (on-robot and eventually off-robot), a stereo camera (on-robot), a 6D

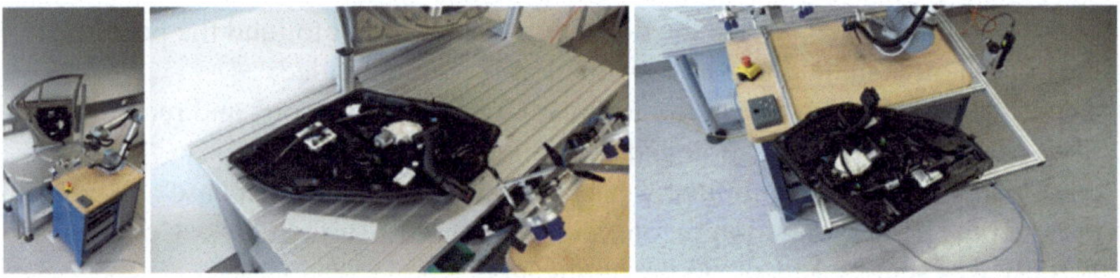

Fig. 2.5 Benchmarking scenarios for Challenge 1

force/torque sensor (on-robot), an automatically attachable screwing tool, an automatically attachable module gripper, a screw tray and a tool changing rack. All controllable components and sensors were connected to a central platform PC running the ROS framework. The Challenger Teams had to connect their PC with the platform PC to get access to the data topics for interacting with the system. A URDF Model of the robot, tools and sensors attached to it were available. The relevant frames were known to teams. The cameras were intrinsically and extrinsically calibrated. The tasks were:

- **Task 1: Pick and insert door module**. This task consisted of 2 sub-tasks, each to be completed within a given execution time. The door module was placed on a table with given positional and rotational tolerances. In the first sub-task, the door module had to be picked up with a suction gripper, in the second sub-task, the module had to be inserted into the corresponding opening of the door.
- **Task 2: Screw door module**. In this task (to be completed within a given execution time), the door module was placed as in previous sub-task, a set of screws were provided in a gridded tray, with a given tolerance, to be picked up one by one by the screwing tool and the attached magnetic bit. The screws had to be picked and inserted into 3 holes in module and door using the automatic screwing tool (at least 4 turns deep to be counted as successful insertions).
- **Task 3: Teach and assemble unknown door**. In this task (to be completed within a given execution time), the benchmarking setup was equipped with an unknown door and door module, with a different geometry with respect to the known ones. For each run the door was placed at a given location, known with a given tolerance. The screws were provided in a gridded tray with a given tolerance and had to be picked up one by one by the screwing tool and the attached magnetic bit. After picking up one screw it had to be inserted into one of 3 holes in module and door. The team had to teach the system to perform a door module assembly (pick up and insert door module, pick up and insert 3 screws at the indicated positions). After the teaching, 3 runs had to be performed, each starting and ending with the robot being in a given reference position.

Tasks 1 and 2 could be performed 3 times, of which the best score for each task was chosen for evaluation.

In **Challenge 2**, the benchmarking tasks were carried out in a laboratory environment (Fig. 2.6) equipped with two similar tables and a shelf for storage of Small

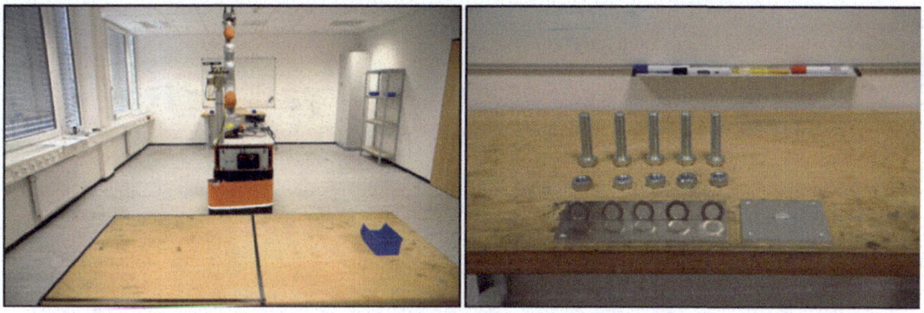

Fig. 2.6 Benchmarking scenario for Challenge 2

Load Carriers (SLCs). A third table could be used for pre-assembly tasks. The setup was composed by an omnidirectional mobile base with safety fields, a 7-DoF robot arm with payload 14 kg (KUKA iiwa), a parallel jaw gripper, a pan tilt unit, a camera system on pole, a camera system at TCP, a server PC (with limited access to contestants), a sensor PC and a power supply. Sensors and cameras were calibrated. Two identical robotic platforms were prepared and made available to Challengers. The software included the ROS framework for all provided hardware components, a navigation system with fine positioning function and a simulator to work with the same ROS interfaces as the real hardware system.

A logistic task and an assembly task had to be solved:

- **Task 1: Bring SLCs loaded with different parts to a workplace**. In this Task, to be completed within a given execution time, 5 SLCs were placed in different locations, while the robot was initially placed in front of the table where the goal location was. The task required to pick and place SLCs with different payloads from different locations and delivery to a fixed place.
- **Task 2: Screwing of nuts and bolts**. This task was composed by two sub-tasks and had to be completed within a given maximum execution time. In the first sub-task, 5 nuts, screws and washers were placed on a table within a fixed area; next to the parts a fixture was mounted in a fixed relation to the table, which gave the head of the screw a fixed position for screwing the nut on. The task was to insert screw into fixture and place washer on screw. The second sub-task adopted the same setup, while the goal was to insert screws into nuts.

In **Challenge 3**, the benchmarking environment was a flying arena (Fig. 2.7) equipped with L-shaped scaffolding (as basis for obstacles, such as narrow passages, open space), a Vicon Mocap motion capture system. The robotic hardware consisted of 5 multi-rotor helicopters (1 per team) equipped with on-board sensors. The arena was equipped with a vehicle monitoring system (maximum attitude angle and velocities, minimum and maximum thrust, proximity to the arena bounds and obstacles). The basic software building blocks for getting the MAV airborne were provided, and were based on ROS: interface to MAV, linear MPC for position and trajectory control (using MAV's internal attitude controller), multi-sensor fusion framework

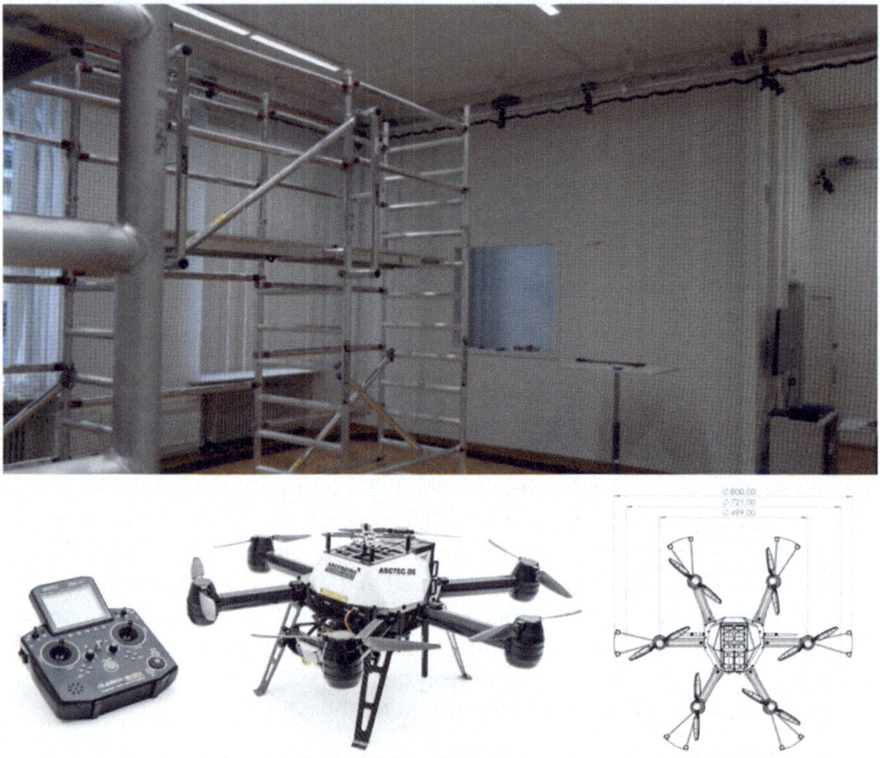

Fig. 2.7 Benchmarking scenario for Challenge 3

(fusing multiple measurements with IMU), localization system using tightly coupled visual-inertial fusion and tools for trajectory generation and sampling.

Five tasks had to be solved:

- **Task 1: System setup**. This task benchmarked the time needed to set up the platform out of the transportation box (i.e., the vehicle was initially packed in the transportation case and all components powered off) and achieve stable hovering.
- **Task 2: State estimation and controller accuracy**. The purpose of this task was to assess the drift of the visual localization system and state estimator, as well as the quality and robustness of the controller. Collision-free random trajectories have been published to the Challenger's system and had to be followed as accurately as possible. The task consisted of five sub-tasks: Sub-tasks 1–4 coincided with the four segments of an assigned trajectory to be tracked (there were no stop between sub-tasks, i.e., the whole trajectory had to be flown at once), the second sub-task was a trajectory following task in the presence of disturbances and turbulences from wind.
- **Task 3: Waypoint navigation**. This task benchmarked local map creation as well as planning. First, the arena had to be explored by following a collision-free trajec-

tory that covered the arena quite well, but not completely. Then, collision-free paths had to be planned to several waypoints with increasing difficulty, and followed by the vehicle. Each waypoint defined a sub-task.
- **Task 4: Reactive obstacle avoidance**. This task benchmarked reactive collision avoidance, useful for semi-autonomous operation, where for instance an unskilled pilot or a pilot from far distance sends velocity commands to the vehicle, but cannot see/assess obstacles. For repeatability, the Challenge Host computer simulated an unskilled pilot by giving joystick (i.e., velocity) commands to the Challenger vehicle, steering it into/towards multiple obstacles on purpose. The task consisted of five sub-tasks, involving obstacles with increasing difficulty: from a well-textured vertical wall, up to obstacles having objects sticking out, hanging cables or dirty glass walls.
- **Task 5: Structure following**. The goal of this task was to cover a specified area in minimal time for inspection of a bent and/or non-vertical wall. The area had to be covered within one single flight.

In some cases relative scoring was adopted, i.e., a number of metrics had to be scored relatively to the highest achieving team. When the metric had to minimized (e.g., time, error), the following formula was adopted

$$\text{score}[\%] = \frac{\text{best metric}}{\text{team's metric}} \times 100\,,$$

while the formula

$$\text{score}[\%] = \frac{\text{team's metric}}{\text{best metric}} \times 100\,,$$

was adopted when the metric had to be maximized.

Evaluation was carried out on the basis of:

- official video footage and data captured during the evaluation runs,
- Challenge Host report with results and scores achieved by the Challenger Team,
- interviews with the Challenger Teams during the **mid-term evaluation event** (Zurich, 19–20 September 2016).

For this round, the role of the Evaluation Board was twofold:

- retrieve the results (scores) obtained by each Challenger Team in the Benchmarking round from the reports prepared by the Challenge Hosts, as well as any other additional information reported;
- ascertain that the activities were carried out in compliance with the rules and procedures laid down by EuRoC.

2.3.2 Free-Style Round

The evaluation runs for the Free-style tasks were carried out on the three EuRoC platforms (see Sect. 1.3.1 for technical details on the platforms), according to a schedule determined by the Challenge Hosts.

Free-style evaluation was based on the following two factors:

- The extent to which the Challengers met measurable goals/metrics determined by each team. To this aim, each team admitted to Stage II was invited to provide a list of objectives, metrics and targets with regard to its free-style task in advance. At least three and a maximum of five objectives for the free-style round had to be provided by each Challenger Team. Objectives were goals the teams wanted to reach within the free-style tasks, which could be quantified with certain metrics and thus progress could be measured and evaluated in an objective way. At least one and a maximum of three metrics per objective had to be provided by each Challenger Team, together with the targeted level of improvement (T) with regard to each metric (absolute or relative) as well as the current baseline (B) for that metric. Thus, the Evaluation Board received 5 documents per Challenge, each collecting objectives (with a brief description), metrics, baseline and targets of each Challenger Team.
- The difficulty of the proposed tasks, measured on a scale determined in advance by the Evaluation Board. In order to achieve a fair and robust determination of the difficulty points to be distributed across all objectives and metrics, each team involved in a Challenge was asked to propose the difficulty points related to each metric for all the 5 teams (including those of the team itself) involved in the same Challenge. Then, the difficulty points proposed by teams were sent to the Evaluation Board, who used them as a reference (but they were not binding in any way) and determined their values definitively. The determination of the difficulty points followed a two-step procedure:

    1. Each board member assigned difficulty points and also stated his/her confidence in his/her assignment.
    2. EuRoC circulated the difficulty points and confidence levels determined in the first step determined by all evaluators (the names of the evaluators were hidden) and asked each board member to re-evaluate his/her assignments in the light of the assignments. Then, the new individual evaluations were collected and their arithmetic averages were computed. Each average was rounded to the closest integer number in a scale ranging from 0 (denoting errors/inconsistencies in the definition of the metric/target/baseline) to 5 (target extremely challenging). The difficulty points determined by arithmetic averaging, after rounding, were then communicated to the Challenger Teams and the Evaluation Board.

  The difference in total difficulty points assigned to the various Challenger Teams was not an expression of the number of metrics put by each team, but of the overall difficulty level of the Free-style task as estimated by the Evaluation Board. Since the difficulty points had to reflect the overall difficulty level of the task, the same

	Target (T)	Baseline (B)	Achievement (A)	Improvement (I) $I = \frac{\lvert(A-B)\rvert}{\lvert(T-B)\rvert}$	Difficulty points (D)	Awarded points (P) $P = I * D$
Objective 1						
Metric 1.1	T1.1	B1.1	A1.1	I1.1	D1.1	P1.1
Metric 1.2	T1.2	B1.2	A1.2	I1.2	D1.2	P1.2
Metric 1.3	T1.3	B1.3	A1.3	I1.3	D1.3	P1.3
Score (obj. 1)						S1=Mean(P1.1,P1.2,P1.3)
...						
Objective N						
Metric N.1	TN.1	BN.1	AN.1	IN.1	DN.1	PN.1
Metric N.2	TN.2	BN.2	AN.2	IN.2	DN.2	PN.2
Score (obj. N)						SN=Mean(PN.1,PN.2)
Total score						S=Mean(S1,...,SN)

Fig. 2.8 Scheme for the computation of the Free-style scoring

metric for different teams could be awarded by different difficulty points, if the level of difficulty of the corresponding objectives was judged different.

For each metric, the progress of the team was measured through the improvement (I), computed via the formulas reported in Fig. 2.8. The improvement was then multiplied with the corresponding difficulty points to obtain the awarded points (P) for the metric. The score assigned to each objective was computed as the mean value of the points awarded to each metric related to the objective; in this way, the score was made independent of the number of metrics. Likewise, the overall score of the team was computed as the mean value of the scores assigned to each objective; in this way, the overall score was made independent of the number of objectives (see Fig. 2.8).

In addition to the total score determined via the above described mechanism, the Evaluation Board could assign an additional score between 0 and 1 based on the overall quality, relevance and potential impact of the work (*coolness factor*).

Evaluation was then carried out on the basis of:

- official video footage and data captured during the demonstration runs,
- Challenge Host reports with results and scores achieved by the Challenger Teams,
- Challenger Teams reports describing the overall activity during the Free-style task,
- interviews with the Challenger Teams carried out during the **mid-term evaluation event** (Zurich, 19–20 September 2016).

It is worth noticing that most of the teams selected their Free-style task close to the Showcase task, although this was not a requirement.

2.3.3 Showcase Round

The evaluation runs for the Showcase round were carried out according to a schedule determined by the Challenge Hosts.

The evaluation was based on the ex post assessment of the same criteria used for evaluation of proposals and selection of teams admitted to Stage II, i.e., scientific and technical quality of the achieved results, efficiency of the implementation of the activities, impact of the results (see Sect. 2.2.2 for further details). In order to help the Evaluation Board to assess the first criterion, the Challenger Teams were required to determine the goals of their Showcase tasks and submit them in advance to EuRoC. Hence, the Evaluation Board assigned scores to the three criteria on the basis of:

- official video footage and data captured during the evaluation runs,
- Challenge Hosts reports describing the overall activity and the results obtained by each Challenger Teams during the Showcase round,
- Challenger Teams reports describing the overall activity during the Showcase round,
- goals of the Showcase task, determined in advance by Challenger Teams,
- interviews of the evaluators with the Challenger Teams carried out during the **Challenge Workshop** (Stuttgart, 15–17 May 2017).

2.3.4 Admission to Stage III

The scores from Benchmarking, Free-style and Showcase rounds had the same weight in the final score for Stage II. To this aim, the score obtained by a team in each of the 3 rounds was standardized by subtracting the mean value of the scores obtained in the same round by all teams and normalizing with respect to the standard deviation. Standardization allowed to get a dimensionless score not dependent on differences in average and variation in the distribution of the scores in the three rounds. The simple mean of the three standard scores was computed to obtain the final score measure of the overall achievement. More in detail, for each team, the standard score in Round X (N = Benchmarking, Free-Style, Showcase) is obtained as:

$$\text{Standard score in round } N = \frac{\text{Score in round } N - \text{Mean score in round } N}{\text{Standard deviation in round } N},$$

while the overall standard score obtained by the team was computed as:

$$\text{Final total score} = \frac{\sum_N \text{Standard score in round } N}{3}.$$

For each Challenge, the 5 Challenger Teams were ranked according to the final total score: the first 2 in the ranking were proposed to EC for admission to Stage III and funding. An annotated ranking was compiled at the end of the panel meeting, together with the ESRs. The ESRs and the panel meeting minutes were forwarded to the EC. Upon approval by the EC, the 6 (2 for each challenge) **Challenge Finalists** were admitted to Stage III.

The Challengers were informed that a request for redress could be submitted to EuRoC. No redress requests were received.

2.4 Evaluation Activities in Stage III

In Stage III (Field Tests), the selected Challenger Finalists were required to develop and run a pilot experiment at the end-user site, aimed at achieving robust and reusable solutions to real industrial needs.

The final evaluation of the 6 Challenge Finalists was performed by a Board of Judges (BoJ), based on the ex post assessment of the same criteria used for evaluation of Round (b) of Stage II, i.e., scientific and technical quality of the achieved results, efficiency of the implementation of the activities, potential impact of the results.

The BoJ was completely independent from the EuRoC core consortium and was composed by the Challenge Advisor at Large, as the technical/scientific stakeholder who has supervised the three challenges in all their various stages and rounds, the Challenge Advisor for European Robotics Industries, as the expert for exploitation of the results of the challenges, along with 5 external stakeholders (a representative from the field of policy making or government, a representative from a large manufacturing company, a representative from an SME, a representative of the entrepreneur/venture capitalist community, and a media/social expert). The first two members were already part of the Challenge Advisory Board, while the 5 external stakeholders were selected by the EuRoC core consortium.

For evaluation purposes, each Challenger Team was required to:

- determine objectives/targets for Stage III and submit them in advance to EuRoC;
- capture video footage and relevant data, showing and archiving their progress;
- prepare a report (Field Tests Report) describing the activities carried out and the results achieved, as well as the extent to which objectives and targets were met; also, a one-page publishable summary about results and achievements was required;
- participate in an on-site evaluation at the end-user site, attended by at least one observer from the Challenge Host; the observers produced a statement of verification of the extent to which they witnessed the claims included in the Field Tests Report demonstrated.

In order to determine the extent to which the three criteria for evaluation were met, the BoJ considered the following materials:

- video footage and data provided by Challenger Teams,
- the objectives/targets set by Challenger Teams,
- the contents of the Field Tests Reports and the related one-page publishable summary,
- the statement from the Challenge Hosts observers verifying claims of the Field Tests Reports,
- interviews with Challenger Teams carried out during the **Final Challenge Workshop** (Napoli, 5–6 April 2018).

Before the Final Challenge Workshop, each member of the BoJ provided a preliminary evaluation of the activities and the results achieved by each Challenger Team, by compiling a preliminary evaluation report. During the panel meeting all the evaluation material, including the outcome of the interviews, was discussed so as to reach a consensus on scores and comments. Then, the ESRs were compiled on the basis of consensus scores and comments. The ESRs and the evaluation meeting minutes were forwarded to the European Commission. After approval by the European Commission, the team proposed as EuRoC Winner was declared. Evaluation and selection results were communicated to the proposers. The proposers were informed that a request for redress could be submitted to EuRoC. No redress requests were received.

The EuRoC Winner has been publicly announced and awarded during the Automatica Fair (Munich, 19–22 June 2018).

2.5 Conclusion

In this Chapter the evaluation activities carried out in the framework of the EuRoC project have been presented, with special emphasis on the innovative procedures and rules adopted along the three stages that could be re-used in future similar initiatives. Among the most innovative contributions developed by EuRoC, it is worth mentioning the evaluation procedures of the Free-style task, which was freely determined by Challengers, and, in particular, the determination of the difficulty points, which heavily involved the Challengers. This procedures was appreciated by Challengers and external evaluators.

Chapter 3
The EuRoC Platforms

Daniel Braun, Markus Achtelik, Christian Ott, Bernd Winkler and Zachary Taylor

Abstract This chapter describes the robotics hardware and software prepared and provided by the EuRoC core consortium. The challenger teams have based their individual solutions based on these provided systems thus equal opportunities have been ensured. Throughout the implementation of EuRoC and the challengers' solutions the platforms have also been amended by technology developed by the core consortium. Further changes and additions made by the challenger teams for adjusting to the selected use case in the pilot installations are also described.

Keywords Robot systems · Mobile manipulators · UAVs · Integration infrastructure.

D. Braun (✉)
KUKA Deutschland GmbH, Zugspitzstraße 140, 86356 Augsburg, Germany
e-mail: Daniel.Braun@kuka.com

M. Achtelik
Ascending Technologies Gmbh, Konrad-Zuse-Bogen 4, 82152 Krailling, Germany
e-mail: markus.achtelik@intel.com

C. Ott
German Aerospace Center (DLR), Institute of Robotics and Mechatronics, Münchnerstrasse 20, 82234 Weßling, Germany
e-mail: christian.ott@dlr.de

B. Winkler
Fraunhofer Institute for Manufacturing Engineering and Automation (IPA), Department for Robot and Assistive Systems, Nobelstraße 12, 70569 Stuttgart, Germany
e-mail: bernd.winkler@ipa.fraunhofer.de

Z. Taylor
Autonomous Systems Lab, ETH Zürich (ETHZ), Leonhardstraße 21, 8092 Zürich, Switzerland
e-mail: ztaylor@ethz.ch

F. Caccavale et al. (eds.), *Bringing Innovative Robotic Technologies from Research Labs to Industrial End-users*, Springer Tracts in Advanced Robotics 136,
https://doi.org/10.1007/978-3-030-34507-5_3

3.1 Introduction

One of the core ideas of the EuRoC project was to provide potential challenger teams with capable and integrated state-of-the-art hardware for realizing their applications, in order to lower the entry barriers for teams providing new subsets of technology for new applications and better performance in these applications. Thus the EuRoC core consortium designed and implemented suitable solutions for providing hardware in all three challenge tracks. As the challenge tracks had varying targeted scenarios, the hardware concepts, access methods and the implementation of the challenges themselves differed as well: while a set of hardware and software modules were offered in Challenge 1, two integrated mobile manipulation systems were designed and implemented in Challenge 2. These two challenges provided the hardware at the challenge hosts' sites during the benchmarking stage of EuRoC. In Challenge 3 a specialized UAV was designed and implemented and each challenger team received their own unit from Stage II on.

This chapter provides an introduction to the provided software and hardware systems, on which the challenger teams based the development and implementation of their applications. As the systems have also evolved during the implementation of the challenges, a brief overview of additions and changes on the platforms for the six finalist applications is provided within this chapter. Further details about the finalists' solutions can be found in Parts I, II and III.

3.2 Challenge 1: Reconfigurable Interactive Manufacturing Cell

3.2.1 Summary of the Provided Systems

The platform provided in Challenge 1 was not a fully integrated system as in the other two Challenges (see Sects. 3.3 and 3.4) but a set of platform components covering a wide range of automation needs. These could be used and have been used by the teams to build their setup during their stay at the hosts site.

From the start of Stage II (Realistic Labs), the platform components summarized in Table 3.1 were offered to the teams at the challenge host site.

In addition to providing the individual components for the Benchmarking round a robotic setup built from the components was provided by the host (Fig. 3.1).

As described in the original challenge design, the platform components were only available to the teams during their stay at the host site. When working at their own premises, the teams had to provide for their own hardware setup. The software interfaces developed and maintained by the host were made available to the teams thus allowing faster development at home and quick integration when being at the host site.

Table 3.1 Equipment offered to the challengers in Challenge 1

Equipment	Count	Software Interface
6-DoF robot (universal robot UR10)	1	ROS
7-DoF robot (KUKA LBR iiwa 7 kg)	1	ROS
6-DoF robot (KUKA KR16-2 + KR-C4)	1	ROS
Depth/RGB camera (Kinect)	2	ROS
Depth camera (pmd technologies CamBoard nano)	1	ROS
Stereo camera (Dual head VRmagic D3)	1	ROS
Force torque sensor (Robotix FT 150)	1	ROS
Passive robot platform	1	n.a.
Passive door platform	1	n.a.
ROS-PC	1	ROS
Light curtains (Sick miniTwin4 1200 mm and 420mm)	2	Digital IO
IO-board (PhidgetInterfaceKit 0/16/16)	2	ROS
Electrical 2-Finger gripper (Schunk WSG50-110)	1	ROS
Various 2 and 3 fingered pneumatic grippers	5	Digital IO
Automatic screwing tool (Weber Pluto 10)	1	Digital IO
Safety PLC (Siemens S7)	1	Digital IO, proprietary

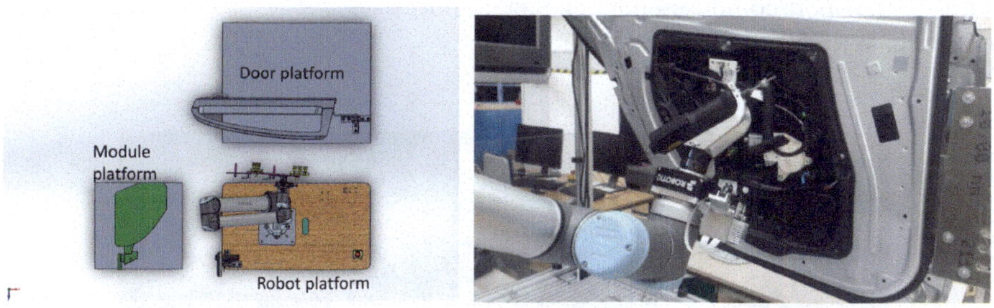

Fig. 3.1 Provided setup (left) and robot system for the Benchmarking round in Challenge 1

3.2.2 Changes and Additions Developed During Stage II

Despite being offered to use the platform components on-site, most teams decided to ship the full setup they used in their own labs to the host site during their stay to avoid potential problems with switching hardware components. Still, on-site several of the components were integrated, e.g. four teams used the universal robot UR10 for demonstrating aspects of their developed solutions or the KUKA LBR iiwa robot was used as backup robot as the one shipped by the team was failing.

Overall the main changes on the platform provided by the host concerned improvements and maintenance of the software drivers of the components. The setup of the teams shipped to the host site for the demonstrations were specifically designed to cover the requirements of the specific task of the team and were hugely different from each other and comprised specialized hardware.

3.2.3 Changes and Additions Introduced Towards Stage III

As discussed in Sect. 3.2, the challenge hardware was provided to the challenger teams on–site at the challenge host, while the teams used their own hardware in their labs. Therefore the teams implemented the robot cells for the pilot experiments in Stage III also with own hardware. The setups are briefly discussed herein to provide a basis for further component and framework development for setting up such cells in future.

During the Stage III team PIROS continued development on their own specialized hardware. No changes implemented by the host were deployed on the setup of the team during this stage. In Stage IIb (Showcase round, see Fig. 3.2) own hardware was deployed, continuing testing both UR10 and KUKA iiwa arms inside the same software framework, and continuing developing end–effectors.

The hardware setup included

- KUKA iiwa R820 (replaced by KUKA iiwa R800 during Stage IIb demonstration, due to transport/commissioning damage/failure)
- Schunk WSG50 Co-Act gripper

Fig. 3.2 PIROS setup during challenge stage IIb at the challenge host, both robot arms were used around the target machine tool pallet (at the bottom)

- custom-design pneumatic gripper for heavy load (based on Schunk PGN+ 160-2)
- custom-design vacuum grippers
- Universal Robots UR10 robotic arm
- Ensenso 3D vision sensor
- stereo-camera onboard system
- Schunk SWK-011 tool changer
- B&R S8000 SafeLogic.

The purpose of Stage IIb, deployment was mainly dedicated to testing software integration of several components, and the modularity of controllers, including the deployment of assembly strategies on different platforms presented in Stage IIa. The software integration included several user interfaces for enabling flexible allocation of reconfigurable collaborative tasks:

- Tablet PC
- Myo armband for gesture recognition (electromyographic signals)
- Microsoft Kinect for gesture recognition (vision).

In Stage III (Field tests), the own hardware was engineered for deployment on the factory floor (see Fig. 3.3).

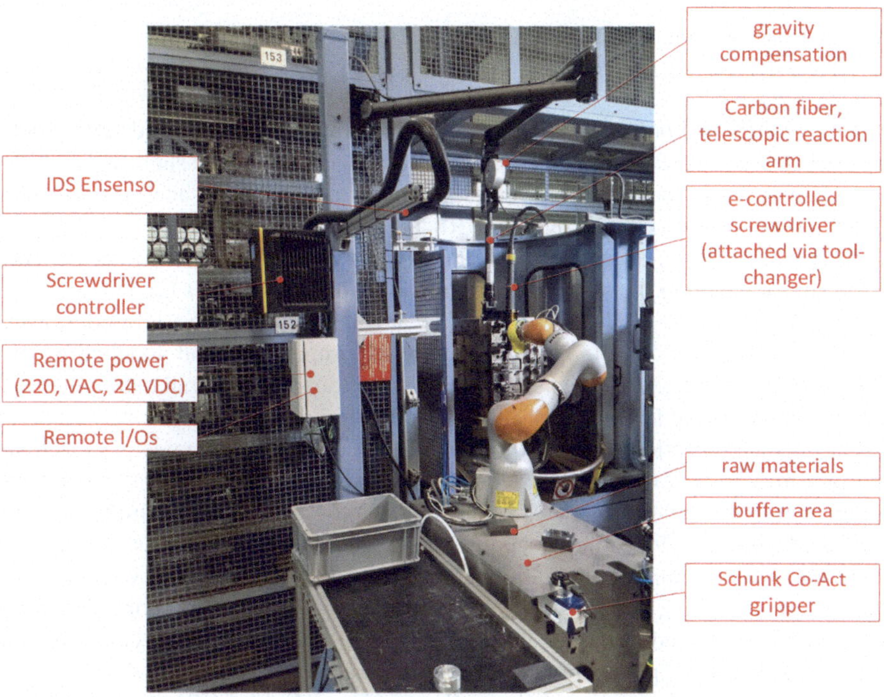

Fig. 3.3 Additional hardware introduced by PIROS team for system integration at shop floor deployment stage (courtesy of Cembre SpA)

Fig. 3.4 Front view of the PIROS system deployed at Cembre SpA for Stage III

All electrical and power components were integrated in cabinets and carts. Tools and tool changer were confirmed from previous stages. Protective shells around on-board cameras and tool changer were added for collaborative usage purpose. The selected robot arm was the KUKA iiwa R820.

As a main feature, an electronic-controlled screwdriver was added to the own hardware for enabling the shared use of the tool by the robot and a human operator. The high torque (up to $90Nm$) is balanced by a reaction carbon-fiber telescopic arm, directly attached to the tool. The tool is gravity-compensated and attached by the robot with the standard tool changer used for other tools. The overall deployment of own hardware on the factory floor is illustrated in Fig. 3.4.

During the Freestyle round FLA^{2}IR team focused on intuitive trajectory teach–in and on force–based exploration, execution and user interaction. A major software component in this phase was FLA^{2}IR's own CAD-2-Path software, which was also used and further developed in succeeding stages. FZI's own UR10 robot was used as manipulator for the final evaluation. In addition, a force-torque sensor was added to the end-effector along with a custom tool for drawing on car doors. For controlling the robot FLA^{2}IR used the public `ur_modern` ROS package and parts of their own framework, because this combination proved the richest feature set.

For the Showcase round and Stage III (see Fig. 3.5) FLA^{2}IR targeted robotic assembly of flexible polymer sealing strips to car doors. Due to this new overall objective, no hardware but only software was re-used from the freestyle phase. In

Fig. 3.5 Final setup prepared by team FLA^{2}IR for StageIII, with the pneumatics-enhanced gripper (left) and the robot in front of a car door conveyor at the end-user's site (right)

addition, both the showcase and the use case where too specific, so that FLA^{2}IR could not build on hardware or software components provided by the challenge host. The following list gives an overview of the major hardware components used during the Showcase round:

- KUKA KR16 robot with custom safety skin
- End-effector force-torque sensor ATI delta
- Industrial gripper Schunk PG70+
- Custom pinch gripper with two jaws and optical fibre
- Mountings for the robot, the car door and a rack for the sealing strips
- SICK laser scanners for the zone-based safety concept.

For the pilot experiments developed in Stage III the overall setting of the showcase was further enhanced. The following list gives an overview of the major improvements for the hardware with respect to the showcase. Like before, no hardware was used from the challenge host.

- Custom pinch gripper in enhanced design with pneumatic piston
- Original car door conveyor at the end-user's site.

The following list presents the major software components used for the pilot experiments:

- FZI's CAD-2-Path: A browser-based application for intuitive trajectory teaching
- FZI's motion pipeline: A framework for teaching, managing and executing complex trajectories
- FZI's Cartesian controllers: A set of ROS controllers for Cartesian end-point following, force and compliance control
- FZI's fork of the `kuka_experimental` ROS industrial package: New features for pausing and stopping robots in a security context
- a hierarchical state machine based on flexbe.

3.2.4 *Summary of Requirements for Future Development*

Overall the teams innovated and implemented their own solutions for many diverse aspects of solving robotic automation tasks. Foundations, however, like device drivers for robots and peripheral components are enablers for those innovations and should be maintained and continuously improved, as well as extended to new robot models that are attractive for the manufacturing/automatic assembly context.

While it is hard to foresee a standardized hardware interface (e.g. standard electrical interfaces for end effectors, including high bandwidth interface for vision), software frameworks can be improved with maintained, up-to-date and bug-fixed, repositories of ROS/ROS-I drivers and bridges for major automation protocols.

The `kuka_experimental` stack of ROS industrial is a very useful baseline for applications similar to the FLA^{2}IR use case. The driver uses the ROS control concepts, which is why it was easy to integrate with existing frameworks. This should hold true for many roboticists. It has been the impression of the teams that ROS control is becoming and has become the dominant "way" of controlling robots. It would be worth pushing this for the still missing robotic manipulators in the field.

3.3 Platform in Challenge 2: Shop Floor Logistics and Manipulation

3.3.1 *Summary of the Provided System*

The challenge hardware and software system originally specified and provided by DLR and KUKA to the Challengers included a mobile manipulation system (Fig. 3.6), consisting of an omni-directional mobile platform with a 7-axis robot manipulator as

Fig. 3.6 Integrated mobile manipulation system provided as platform in Challenge 2

Work space

Dimensions (mm)	A	B	C	D	E	F	G	H	I	Volume
LBR iiwa 14 R820	1.291	1.180	360	420	400	250	171	85,5	820	1,8 m³

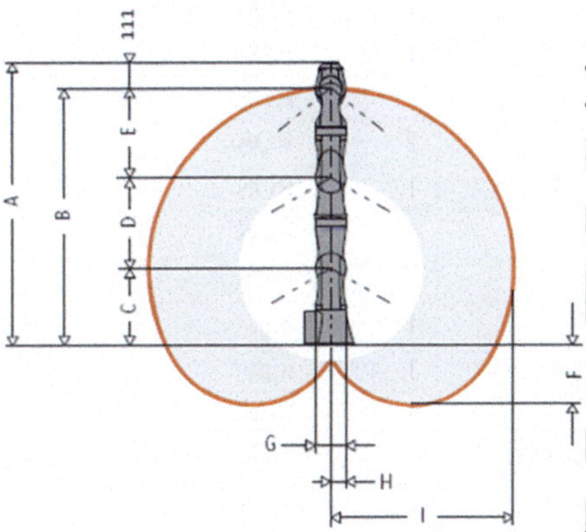

LBR iiwa	14 R820
Rated payload	14 kg
Number of axis	7
Hand variation	Main hand
Mounting flange A/7	DIN ISO 9409-1-A50
Installation position	bottom
Repeatability (ISO 9283)	± 0,1 mm
Torque accuracy axis specific (from the maximum torque)	± 2%
Weight	29,5 kg
Safety	IP 54
Control	KUKA Sunrise Cabinet

Fig. 3.7 KUKA LBR iiwa work space dimensions and mechanical specifications

well as additional sensor and actor hardware. For the operation of the system the basic functionalities for programming and operating manipulator arm and mobile platform were provided by KUKA, while higher level functionality, additional computing facilities and interfaces making these capabilities accessible have been provided by the Challenge Host DLR.

The mobile manipulation system consisted of the sensitive torque-controlled 7-axis robot manipulator KUKA LBR iiwa. The variant featuring 14 kg payload and a total reach of 820 mm was used for the EuRoC mobile manipulator (work space specification and mechanical specifications are characterized in Fig. 3.7).

The robot manipulator was the same unit as the stand-alone robot commercially available from KUKA. This also included the specifications for HRC (human robot collaboration). For maximum flexibility, a flange version with electrical wiring (power and data) was chosen. With the fully integrated wiring of the robot and accessory connections collaboration and autonomous motion planning scenarios could be addressed much more easily.

The robot controller cabinet was modified to control both, the robot arm and the mobile platform. It also integrated a dedicated computer which provides the autonomous navigation functionality of the mobile manipulator. The operation of the complete system was still provided using the KUKA teach pendant (KUKA smartPAD). Further extensions include a safe remote controller for the mobile base vehicle with emergency stop for the complete integrated system.

The mobile base itself (for specifications see Table 3.2) was an omni-directional vehicle based on four individually actuated and controlled Mecanum wheels. The vehicle was fully autonomous and carries power supply for around 8 h independent

Table 3.2 Equipment offered to the challengers in Challenge 1

Equipment	Count	Software Interface
6-DoF robot (universal robot UR10)	1	ROS
7-DoF robot (KUKA LBR iiwa 7 kg)	1	ROS
6-DoF robot (KUKA KR16-2 + KR-C4)	1	ROS
Depth/RGB camera (Kinect)	2	ROS
Depth camera (pmd technologies CamBoard nano)	1	ROS
Stereo camera (Dual head VRmagic D3)	1	ROS
Force torque sensor (Robotix FT 150)	1	ROS
Passive robot platform	1	n.a.
Passive door platform	1	n.a.
ROS-PC	1	ROS
Light curtains (Sick miniTwin4 1200 mm and 420mm)	2	Digital IO
IO-board (PhidgetInterfaceKit 0/16/16)	2	ROS
Electrical 2-Finger gripper (Schunk WSG50-110)	1	ROS
Various 2 and 3 fingered pneumatic grippers	5	Digital IO
Automatic screwing tool (Weber Pluto 10)	1	Digital IO
Safety PLC (Siemens S7)	1	Digital IO, proprietary

operation, robot/platform controller and the robot manipulator arm. Integrated laser scanners provide safety features for the complete system when sharing the same work space with humans. The laser scanners also provided the system with input data for the self-localization and navigation system allowing the system move autonomously in changing environments without the use of additional markers or external localization systems.

On top of the base vehicle the following additional hardware was integrated for providing sensor data input and allowing the mobile manipulation system to interact with the environment, while solving the application specific tasks:

- Pan-tilt unit on top of a sensor pole
  - stereo camera system 3D imaging.
- Tool
  - 2-finger gripper with force control (via motor current)
  - camera system mounted on the flange.
- Challenger module
  - Power Converter (400W) for powering additional laptops
  - up to 3 contestant laptops (for software and functionality integration)
  - 5 port Ethernet switch
  - power outlet for laptops and small equipment.

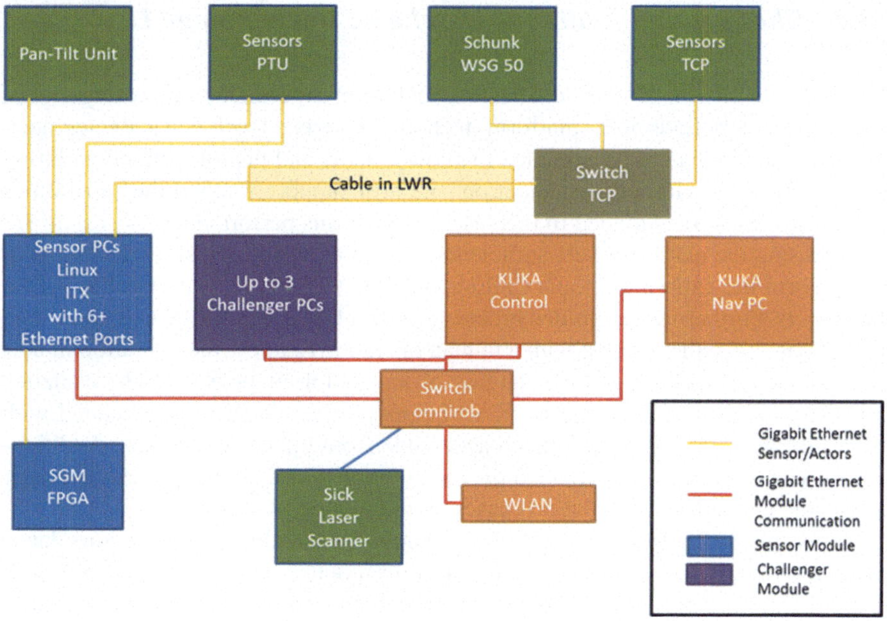

Fig. 3.8 KUKA-DLR communication infrastructure on the mobile manipulator

- Sensor module
  - 2 ITX boards, one powerful graphics adaptor for calculations, 4port network adapter
  - 3 x 5 port switches
  - FPGA board for data processing.
- Power module
  - DC/DC converters from 48 V to 5V, 12V and 24V
  - power distribution and fuses for protection.

The various systems were connected in a network structure (Fig. 3.8) making all components accessible for the provided control computers and also provide connection and communication end points for client side development and control computers used by the Challengers.

On the software side, the provided system included all standard functionality of the LBR iiwa, including the facilities for implementing human-robot-collaboration and interfaces for more direct motion generation. These functionality was augmented by interfaces towards the KUKA.NavigationSolution controlling the behaviour of the mobile base vehicle. On top of the API of the hardware system, the challenge host DLR provided access to more advanced functionality making the sensors and actors accessible and providing support for the setup of complex mobile manipulation applications. This functionality was provided via ROS interfaces.

3.3.2 Changes and Additions Developed During Stage II

During the Realistic labs stage and the Stage III (pilot experiments) two categories of adaptions and additions of the platforms technologies were developed and integrated: First the platform had to be extended to allow reliable and flexible operation during pilot experiments without permanent on–site host support. Second the capabilities of the platform were extended to increase the platforms performance.

Increasing the platforms self-sufficiency: To calibrate and register the sensors of the platform the calibration tools CalLab and CalDe were made accessible for the challengers. Furthermore, a guided process was developed to collect the required data semi–automatic collaborative with a human operator. To allow the operation in new environments, recording and exchanging of navigation maps was made available. This process was based on the KUKA Mapping Tools which were extended with a converter to make the visualization accessible from the ROS interface. Additionally, the user options were extended to configuration and adapt the process manager initializing and configuring the processes on the robotic platform.

Increasing the platforms capabilities: To support a user during the low impedance mode, a self–collision avoidance and joint limit monitor based on the KUKA connectivity interface was developed and integrated. To analyze the robot's workspace, capability maps for the platform were created and delivered. Based on these maps, the manipulability of objects in the robot's workspace can be estimated. A generic object localization tool was provided as well. This tool localizes objects based on the sensor data acquired by the platform. To control a complex robotic system, multiple components have to be coordinated and configured. The traditional sequential programming of complex robotic application is a challenge. In the last years state machine based application programming approaches delivered promising results. Therefore a graphical programming tool (RAFCON), based on the state machine concept, was developed and provided onboard.

3.3.3 Changes and Additions Introduced Towards Stage III

The RSAII team mainly adjusted the environment of the provided mobile manipulation system in order to fit the targeted scenario. The mobile platform was adapted to perform logistic operations, specifically unitary order preparation and return tasks. To achieve this objective, several hardware and software elements were added:

- a conveyor belt layout with several transfer units, containers and cameras (Fig. 3.9)
- the robot arm of the mobile manipulation system was outfitted with a tool for object picking. Here, the tool was integrated with the robotic arm and used an electric suction system, controlled from a small input/output control box on the mobile platform (Fig. 3.10).

From the point of view of the hardware, TIMAIRIS field tests used the platform in a very efficient way, as the modifications to existing resources were kept to the

Fig. 3.9 Conveyor layout for moving product containers designed and implemented by the RSAII team for unitary order preparation and return tasks

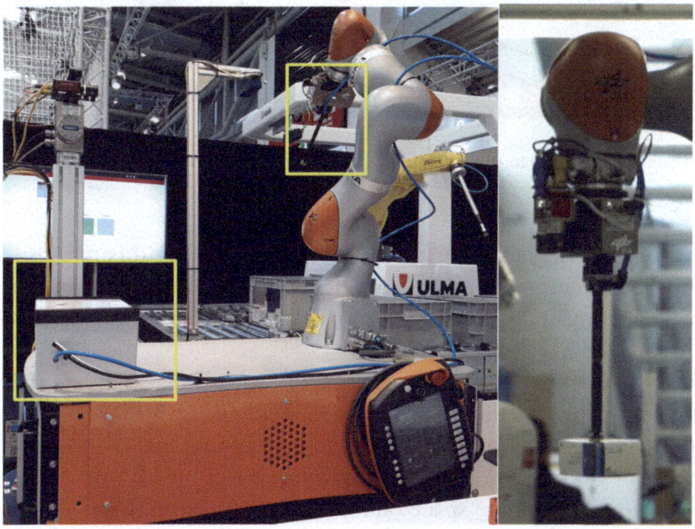

Fig. 3.10 Tool and control box for picking objects implemented on the EuRoC mobile manipulation system by the RSAII team

minimum. A blank gripper was designed and constructed by joining a Zimmer 2-Jaw Parallel Gripper with Long Stroke (Series GEH6000IL) with mechanical parts (fingers) developed by IMA with shape adapted to the blank piles. This gripper was used to grasp 15 cm high blank piles, that would be impossible to grasp with the SCHUNK gripper provided in the platform, due to lack of force. The Zimmer gripper works as an IO-Link slave. To make it work, an IO-link Master is needed to communicate with it. Since the Sunrise cabinet can work only as a Profinet/Ethercat Master, a Gateway

Fig. 3.11 TIMAIRIS challenge platform for autonomous blank feeding

is needed (Profinet/Ethercat slave and IO-Link Master). For this purpose, a Balluff Gateway has been used. A driver was developed to enable communication with the Zimmer gripper using ROS. Figure 3.11 provides some details on the blank gripper configuration.

For human detection, two low–cost Hokuyo Laser Range Finders (LRFs) were mounted on both sides of the platform manipulator (Fig. 3.11). Human detection was used to increase safety of operations and to enable improved forms of human-interaction, as for example, gesture recognition.

In addition, in the available space on top of the mobile platform, a buffer to hold a pile of blanks and a platform to flip the blank piles were added. Each of these consisted in several aluminum parts. Both devices worked together with the new gripper to make the robot more effective by transporting large piles and have the option of transporting two piles simultaneously.

Apart from the mentioned changes, the platform was not modified: the single arm was used to provide all the manipulation, the moving capabilities to provide the navigation and the already existing cameras to provide the perception. The ASUS Xtion, provided by the Challenge Host, and used in the Benchmarking and Freestyle tasks only as a projector to increase the texture of the environment, was used for gesture recognition and thus was directly connected to a challenger laptop.

3.3.4 *Summary of Requirements for Future Development*

During the implementation of Challenge 2, the following issues and recommendations for further development of the challenge platform can be formulated:

- The safety functionality of the integrated laser scanners should not be limited to emergency stop functionality. Intelligent speed overrides or in-application signalling of violated warning fields would enhance the efficiency of executed applications. The current laser scanners do not support dynamic changes of the scanning fields.
- ROS interfaces to the platform and manipulator functionalities would make integration much faster and easier
- Integrated online re-planning functionalities for collision avoidance of the robot manipulator would be a great benefit for the implementation and execution of applications in changing and dynamic environments.

In general, the stability and speed of execution of the used algorithms were a subject for technology development in the finalist applications and the general technological field of mobile mainpulation.

3.4 Platform in Challenge 3: Plant Servicing and Inspection

3.4.1 *Summary of the Provided Systems*

At the start of the Realistic labs stage each challenger team was provided with a prototype Asctec Neo hexacopter, shown in Fig. 3.12.

These systems had the following specifications:

- complete redundancy in all vital components (trinity flight controller, motors, batteries, etc) to protect against a single point of failure
- flight times of up to 26 minutes
- 2 kg payload
- 9 inch propellers

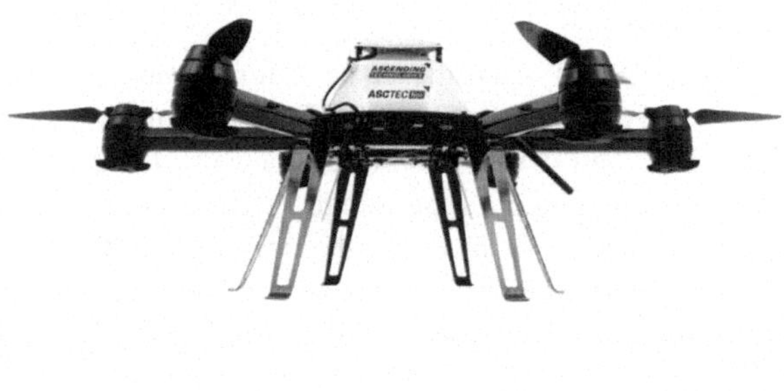

Fig. 3.12 Asctec New Hexacopter as provided by the technology provider and challenge host in C3

- 5 kg maximum takeoff weight
- foldable propellers and detachable arms for easy transportation.

The system was augmented with an onboard Intel-NUC with Core i7-5557U for computation. Sensing in the form of a state-of-the-art visual inertial sensor provided by Skybotics was also installed.

3.4.2 Changes and Additions Developed During Stage II

During the Realistic labs stage teams were provided with updates to existing software tools as well as the addition of some new ones. This included a refactored and updated version of `MAV_trajectory_generation`, the addition of outlier rejection to `ethzasl_msf` and the creation of an `image_undistortion` library for performing a large range of image undistortion, rectification and stereo imaging tasks.

On the host side, only one hardware modification was made to the platforms during the Realistic lab stage. Due to the short lifetime of MAV batteries many systems were beginning to show reduced flight time. Because of this Intel provided an alternative battery interface that allowed connecting Intel Falcon batteries to the systems.

On the Challengers side, the teams were free to augment the hardware of the system. Most teams at this stage chose to add additional depth sensing capabilities among other sensors and actuators. As all teams were provided with individual hardware, a variety of modifications was implemented by the teams, in order to fit the systems to the needs of the individual applications.

3.4.3 Changes and Additions Introduced Towards Stage III

In Stage II the TUM Flyers team used the MAV platform provided by the challenge host, that already czme with a Skybotix VI-sensor supporting 4 cameras with a wide field-of-view (approx. 122°), two of them in a stereo setup in frontal-downward direction. Additionally, an RGB-D camera (Asus Xtion) and 10 ultrasonic distance sensors (Devantech SRF02) were mounted on the drone to cover blind spots of the cameras and to increase the variety of perceivable obstacles (e.g. transparent, poorly illuminated surfaces).

For the Field tests, the Asus Xtion camera was replaced by an Intel Realsense R200, which also provided RGB-D measurements in outdoor environments (see Fig. 3.13). To operate the system, a van of the end user was adapted to place an operator laptop and a long–range WiFi antenna system inside. Those modifications allowed the team to successfully carry out the Field tests.

The GRVC-CATEC team carried out the work in Stage II using the original hardware and software provided by the Challenge Host. The only additional element

Fig. 3.13 Modified MAV used by team TUM Flyers during the Stage III

Fig. 3.14 Setup of the GRVC-CATEC team: RGB-D sensors on-board the MAV, one at the front and another one on the back (left), signalling lights have also been added (right)

was a radio sensor installed on top of the platform, connected by USB to the on-board computer. This sensor was used for achieving two objectives in the Freestyle task: radio-based localization of items and long-term navigation using Range-Only SLAM.

For the Showcase round, the system hardware was modified in different ways, according to the use case in logistic processes inside an aircraft manufacturing plant. The VI-Sensor, which was the main source of information for vision–based navigation, was replaced by an RGB-D camera. Using this sensor allowed to obtain further and more reliable depth estimations while working indoors, which was the case in the application scenario. Moreover, a second RGB-D camera facing backwards was installed to improve the MAV situational awareness for obstacle detection and avoidance, since originally, only the scene in front of the MAV could be sensed (Fig. 3.14).

To increase the awareness of the workers in the aircraft manufacturing plant that a MAV is in operation, a series of LED lights were installed on the aerial platform. Blinking red lights were placed on top (Fig. 3.14), while fixed red lights were below each of the two front arms.

Another modification in the general navigation system is the integration of range measurements to increase the robustness of the MAV localization estimation for long–term operation. A Time of Flight (ToF) range sensor was installed on-board the MAV to receive distance measurements with respect to three other range sensors that were installed in the testing area. Since the pilot experiments developed in Stage III included the logistic process of light goods in a manufacturing plant, a pick-up and delivery system was designed and installed on the MAV. A small cargo hold was designed and 3D–printed to accommodate small goods for aerial transportation, with two automated doors to release its contents.

The software pipeline relied on the same modules that the Challenge Host initially provided for low-level control. Modifications were made to accommodate the new sensors applied to the software modules, which had to be developed earlier in Stage IIa, for example sensor–based odometry, trajectory planning or obstacle avoidance functionalities.

For Stage III, the team decided to change completely the platform provided by the Challenge Host. Due to the environment of the end-user site, and the expected flying height for the MAV, the combination of sensors (an on-board RGB-D camera and a set of fixed UWB beacons around the site) that were successfully used during Stage IIb did not provide the means for robust and safe operations. The operating area counted with big empty spaces, where objects were usually far away, while the maximum range at which the RGB-D camera provides depth measurements is only 10 meters. On the other hand, the use of UWB beacons implies the installation of infrastructure in the plant that needs to be maintained. Apart from this, the aim was to provide the end user with a system working reliably and safely with the minimum possible setup. This motivated the adoption of a new on-board sensor, a single 2D LIDAR with 30 meters range for both horizontal and vertical localization.

Regarding the MAV platform, instead of using the original AscTec Neo a frame from DJI was used to gain more support and availability of parts. The standard Flame Wheel ARF KIT in its hexacopter version (F550) was the chosen platform, which offers a good trade off between size, payload capacity and flight autonomy (Fig. 3.15). The autopilot for low-level control was changed to the popular Pixhawk, making use of its PX4 firmware, and the computer in which the on-board processing tasks were carried out was an UP board.

With respect to software modifications, ROS was maintained as the main programming framework, updating it to its kinetic version in computers running Ubuntu 16.04. The software pipeline was very similar to the one provided by the Challenge Host, but obviously, the low-level controller was switched over to the Pixhawk platform, and a whole new architecture for accommodating the new sensing approach based on LIDAR was developed.

Fig. 3.15 New aerial robot for the realization of the GRVC-CATEC use case in Stage III

3.4.4 Summary of Requirements for Future Development

The following recommendation and requirements for further development of the challenge platforms towards components for solving inspection and servicing tasks can be formulated:

- Increase robustness of the flight controller, full restarts were sometimes necessary after several unsuccessful attempts to take off.
- The structure of the battery sockets needs to be improved as well as the mechanism for extracting the batteries (the exchange of batteries without turning off the whole system was very nice though).
- Improve support and availability of parts.

In total the provided systems were found to be capable of handling the tasks, the derived requirements can mainly be attributed to the research and prototype nature of the provided systems.

3.5 Conclusion

The overall view on the state of the art of the challenge platforms at the end of the project provides a diverse picture. While the base system provided in Challenge 2 was mainly augmented by sensors and actors during the Stage II and also for the implementation of the Field Tests, the systems in Challenge 3 were used and augmented by the Challengers in Stage II. During the implementation of pilot experiments in Stage III, one of the team opted to use a completely different system for the implementation. The picture in Challnge 1 is an extension of this situation: even during Stage II the teams brought in their own robotic hardware, while the implementation of the pilot experiments in Stage III depended widely on their own hardware and software. Also, the recommendations for future development work are quite diverse: Challenge 1 and Challenge 3 mainly concentrate on the requirements to provide standardized interfaces and on increasing the stability of provided hardware and software. In Challenge 2 very concrete advices for improving the versatility of the provided mobile manipulation system have been derived from the experience during the implementation.

The reasons for this diverse picture should be sought for in the combination of complexity of the provided system and the targeted solution. The field of applications in Challenge 1 was very wide and the two applications developed by Challenge Finalists were differing quite a lot. While the provided systems and software could be used for kick-starting the implementations, the final applications were implemented and engineered to provide the best solutions to the use cases. The teams in Challenge 1 also made heavy use of their own existing frameworks.

In Challenge 2 the provided system was already a complex hardware and software solution, which had to be used for implementing the benchmarking tasks in Stage IIa. Based on the this phase, the teams also used the provided system and augmented it with sensors and actors. The great number of recommendations for future work mainly results from the intensive work with this provided systems. It should also be kept in mind, that comparable systems are not easily available on the market or at the teams' own premises.

Finally, in Challenge 3 the teams were provided with the hardware also for their own developments outside the Challenge Host site. The reasons for the teams choosing to use different systems can be found in the specific requirements of the individual use cases (as outlined above). It should also be kept in mind, that investment-wise aerial robotic systems and mobile manipulators, such as in Challenge 2, are in a completely different range. This may also be seen as a decision factor for using provided or own hardware.

In total the implementation of the Challenges was not only a success for the teams or potential customers of the solutions, but also for the Challenge Hosts, Technology Developers and System Integrators. A lot of feedback and experience with the systems could be gathered, which will influence future research and development. As a very concrete evidence for the use of the feedback, the KUKA KMP 200 can be used – this mobile manipulator product is the next step after the prototypical KMR iiwa system. It is now offered by KUKA as a product and among own development it builds also on the experience with the KMR made in EuRoC.

Part II
Achievements in Challenge 1: Reconfigurable Interactive Manufacturing Cell

Introduction

Bernd Winkler

The motivation behind Challenge 1 was to direct the innovative power of Europe's leading research teams towards the development of innovative manufacturing solutions covering aspects as diverse as sensor guided assembly processes, flexible parts manipulation, reconfigurable safety, close human–robot collaboration or human–robot task assignment.

While the setup and the tasks of the initial phases of the Challenge—Stage I (Simulation contest) and Stage II (Benchmarking round)—were defined by the Challenge Host to allow a simple and standardized assessment and comparison of the capabilities of the teams, the application scenarios for the later phases of the Challenge—Stage II (Freestyle and Showcase round)—were free to be defined by the teams.

As intended and enabled by the consortium through the matchmaking phase in Stage I, the chosen application scenario was derived by matching the interest and problem space of the industrial end users with the capabilities of the research teams.

Over the course of the matchmaking, highly diverse proposals for application scenarios emerged, ranging from production of hollow-core chocolate sweets, over winding of stator coils to automotive door seal assembly.

While all teams—forged by combining research institution, end user and system integrator during Stage II—put their best effort into attacking the diverse and specific challenges of their end user's application scenario with the capabilities and expertise of the team, the overall challenge was designed that only two teams (Challenge Finalists) per challenge could reach the end.

The teams that had to leave the challenge after Stage II were not eligible for further funding within EuRoC. However, as they already made significant progress towards solving the application scenario of the end user and were matched with a system integrator for tech transfer working alongside them, these teams already had a very good head start into actually providing an industrial solution to their end user, even without being funded through EuRoC.

The two teams that successfully made it to the final stage were Team FLA^{2}IR and Team PIROS.

The application scenario of Team PIROS addressed machine tending by a safe human-aware robot assistant, which is able to pick parts to be machined, insert them into a fixture, tighten the fixture, and remove and inspect the machined parts. At the same time, each sub-task is assignable either to the human operator or the robot depending on the requirements on the shop floor. Furthermore, depending on the scheduled tasks, a risk assessment is carried out automatically which allows discovering—and in consequence covering—potential safety issues with the actual configuration. The functionality of the system was demonstrated at the end of Stage III at the shop floor of the end user Cembre S.p.A. in Brescia, Italy.

The application scenario of Team FLA^{2}IR addressed the automatic force-controlled assembly of a closed-loop car door sealing. For the handling and the assembly of the sealing, the team developed a custom gripper-based tool, which allowed gripping the sealing, inserting the pins of the sealing in the door and the quality control of the insertion at the same time in a compact design. In addition, the team implemented a web-based 3D interface for efficiently and intuitively defining the assembly process. The functionality of the system was demonstrated at the end of Stage III at the shop floor of the end user Opel Automobile GmbH in Rüsselsheim, Germany.

While both teams consequently used their technical skills for addressing the myriad of details of such innovative technical solutions, ranging from hardware design and manufacturing to system and software design, algorithm development, on to efficiently iterating and testing the final solution, one key skill that has been observed in both teams was their pronounced team spirit and enthusiasm which repeatedly helped overcome difficult situations like broken or malfunctioning robots or other potential dead ends.

Chapter 4
PIROS: Cooperative, Safe and Reconfigurable Robotic Companion for CNC Pallets Load/Unload Stations

Federico Vicentini, Nicola Pedrocchi, Manuel Beschi, Matteo Giussani, Niccolò Iannacci, Paolo Magnoni, Stefania Pellegrinelli, Loris Roveda, Enrico Villagrossi, Mehrnoosh Askarpour, Inaki Maurtua, Alberto Tellaeche, Francesco Becchi, Giovanni Stellin and Giuseppe Fogliazza

Abstract Handling and assembling applications with small batch size and high production mix require requires high adaptability, reconfigurability and flexibility. Thus, human-robot collaboration could be an effective solution to ensure production performance and operator satisfaction. This scenario requires human-awareness in different levels of the software framework, from the robot control to the task planning. The goal is to assign high added value activities to the human as much as possible, while the robot has to be able to substitute the human when needed. Team PIROS faces this goal by designing a IEC 61499/ROS-based architecture which integrate safety assessment, advanced force control, human-aware motion planning, gesture recognition, and task scheduling.

Keywords Human-robot collaboration · Manipulation · Human awareness

---

F. Vicentini · N. Pedrocchi · M. Beschi (✉) · M. Giussani · N. Iannacci · P. Magnoni · S. Pellegrinelli · L. Roveda · E. Villagrossi · M. Askarpour
National Research Council of Italy (CNR-STIIMA), Milan, Italy
e-mail: manuel.beschi@stiima.cnr.it

F. Vicentini
e-mail: federico.vicentini@stiima.cnr.it

M. Askarpour
Politecnico di Milano, Milan, Italy

I. Maurtua · A. Tellaeche
Fundation Tekniker, Eibar, Spain

F. Becchi · G. Stellin
Telerobots Labs, Genoa, Italy

G. Fogliazza
MCM SpA, Vigolzone (PC), Italy

F. Caccavale et al. (eds.), *Bringing Innovative Robotic Technologies from Research Labs to Industrial End-users*, Springer Tracts in Advanced Robotics 136,
https://doi.org/10.1007/978-3-030-34507-5_4

4.1 Introduction

Handling and assembling applications with small batch size and high production mix requires high adaptability, reconfigurability and flexibility. Thus, human-robot collaboration could be an effective solution to ensure production performance and operator satisfaction. This scenario requires human-awareness in different level of the software framework, from the robot control to the task planning. The goal is to assign high added value activities to the human as much as possible, while the robot has to be able to substitute the human when needed. Team PIROS faces this goal by designing a IEC 61499/ROS-based architecture which integrates safety assessment, advanced force control, human-aware motion planning, gesture recognition and task scheduling.

4.2 Use Case Description and Motivation

Team PIROS activities deal with the use of collaborative robotics in handling and assembly applications, in the domain of advanced manufacturing and specifically in the use of Flexible Manufacturing Systems (FMS) for machining of metal parts. FMS is a highly-automated machining system, in which production is dynamically optimized and balanced depending on changing conditions (production mix, batch sizes, availability of machines, etc.). Nonetheless, manual labour is still a relevant component in the preparation/verification steps of the machining processes. FMS is a relevant scenario for hybrid human-robot production systems and collaborative robotics in manufacturing, because of the need of both collaborative assembly tasks and workspace sharing conditions along variable tasks.

4.2.1 Industrial Practice

In conventional FMS, operators are mostly dedicated to interfacing with CNC consoles and to completely handling all material flows. Raw materials are assembled into CNC pallets with fixtures, then machined/finished parts are removed, checked and re-assembled for further machining or prepared for shipping. All these operations happen in dedicated parts of the FMS layout, each one called Load/Unload Station (LUS, see Fig. 4.1). In highly automated FMS, pallets are stored inside a (remarkable large) buffer system, and shuttled to eventual machining destinations according to optimized schedules.

A typical situation is shown in Fig. 4.1, while Fig. 4.2 shows the adoption of the PIROS collaborative robotics solution in support of an existing, fully manual LUS.

Examples of heavy workload tasks that benefit from the use of collaborative robots in a LUS include: setting fixtures, mounting/dismounting workpieces into fixtures

Fig. 4.1 Production scenario with heavy-duty manual tasks at LUSs (Courtesy of: Cembre SpA, Brescia, Italy)

Fig. 4.2 A LUS upgraded with a robot system for a collaborative version of pallet preparation tasks: the robot system provides services from lateral approach (from left side), sharing the workspace with a single LUS operator (Cortesy of: Cembre SpA, Brescia, Italy)

before/after machining, inspecting the premachining setups, inspecting the quality of machined parts. In these cases, robots can provide a number of assistive tasks such as kitting parts and tools, handling parts, supporting manual assembly, moving sensors for inspections, etc., according to *nominal* task plans or inline *alternatives* requested by operators.

While collaborative applications and shared spaces are popular production approaches [1], their efficacy and efficiency are critically dependent on the performance of individual sub-task and on the design of such sub-task allocation between humans and robots. Specifically, in a scenario of remarkable variability of parts and processes, the design and allocation of roles and actions between humans and robots may be the key for determining the target efficiency when unfeasible/inefficient tasks are identified and allocated in an optimal way.

In the proposed scenario, collaborative assembly is investigated and evaluated in terms of two principal factors: (i) the ability in performing the sub-tasks, (ii) the efficient sub-tasks allocation to the human and the robots.

4.2.2 Objectives

For sake of clarity, preliminary definitions are reported below (Fig 4.3):

- **variability** in manufacturing is intended here as requested changes in mixed parts, process/assembly sequences, tools used, target timing, changes in design features or materials, during a production plan;
- **flexibility** in manufacturing [2, 3] is the ability to maintain an efficient production goal, or at least similar operational capabilities, in presence of variability without significant modifications of setup or programs (see a review of the evolution of literature about flexibility in [4] and citations therein);
- **reconfigurability** in manufacturing is the ability to attain the same previous objectives but, unlike flexibility [5], undergoing some system changes when production/process exhibits some variability. Attitude for changes is not built-in, needs down-time to reset but can be more cost-effective depending on the size and quality of variations;
- **adaptability** in robotics is intended here as the possibility for a robot system (hardware and software) to be changed by users. As such, the robot system has to provide some built-in flexibility in terms of task allocation, pattern selection, tooling, etc. **Adaptivity**, instead, indicates the ability to automatically react to changing conditions and fit to new behaviors/targets.

In PIROS use case experience, the most important property for the proposed collaborative robotics solution is *adaptability* of the platform to production and usage needs. In particular, in FMS manual tasks the following conditions hold.

- Process flexibility: the FMS enables its hosting shopfloor to afford a variable production mix in the magnitude of 10^3 parts/year, with batch sizes ranging in

Fig. 4.3 Example of FMS pallet, with bolts/nuts jigs, inside a LUS during an intermediate machining step: array of parts on left side need to be disassembled and flipped to the right side array jigs for second machining step. Additional different jigs are present on lateral mounting surfaces of the pallet (Courtesy of: Cembre SpA, Brescia, Italy)

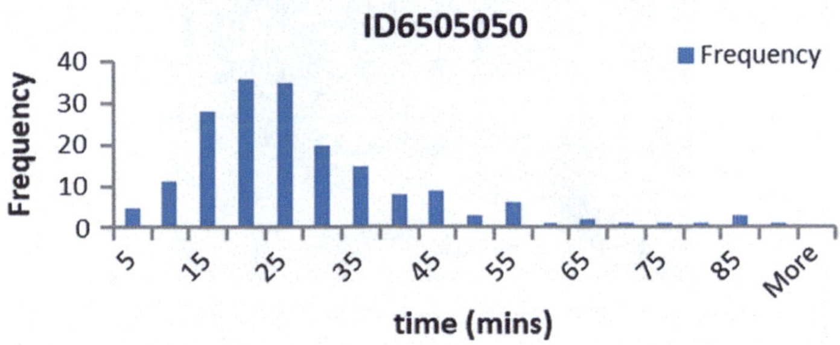

Fig. 4.4 Example of LUS usage profile

$10^2 - 10^5$ parts. Machining pallets have unique fixtures to be prepared, manipulated, checked, cleaned, etc. approximately every 20 min (see an average LUS usage profile in Fig. 4.4). Due to such a production mix profile, the area of the plant integrating the PIROS demonstrator is suitable for reconfigurable solutions. Therefore, additional modules need to cope with the different expected tasks to be performed.

- Process variability determined by human operators: their actions are little repeatable in terms of timing and pace. While assembly/disassembly tasks for CNC pal-

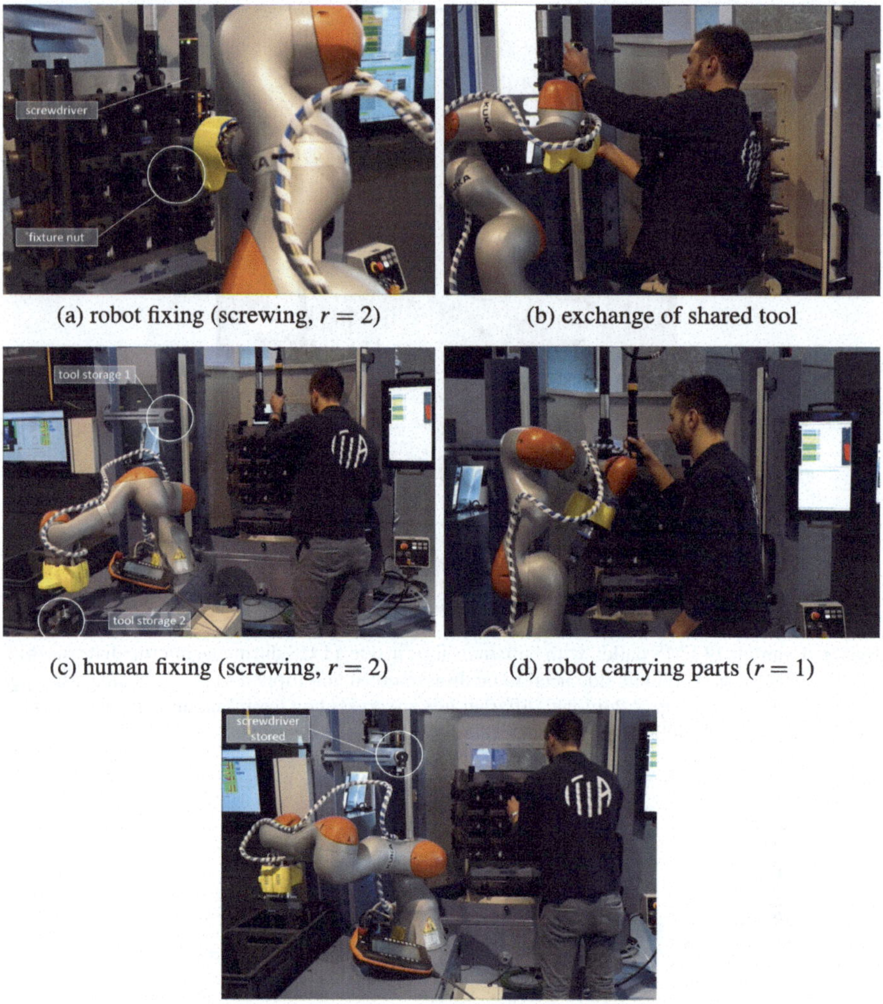

(a) robot fixing (screwing, $r = 2$)

(b) exchange of shared tool

(c) human fixing (screwing, $r = 2$)

(d) robot carrying parts ($r = 1$)

(e) robot bin picking ($r = 1$), human inspecting ($r = 3$)

Fig. 4.5 Example of mixed workflow during an assembly/fixing sub-task (EuRoC field test stage)

lets at LUSs feature standardized operations, we observed operators doing many intermediate variants or alternate usage of time slots, frequently moving in and out of the workstation (e.g., go for materials, go for tools, consult machining CADs, measure, make quality control before resuming, etc). Due to such usage profile, a collaborative robotic solution needs to consider a remarkably large state space of possible task execution modes/sequences, in terms of both sub-programs scheduling and safety conditions.

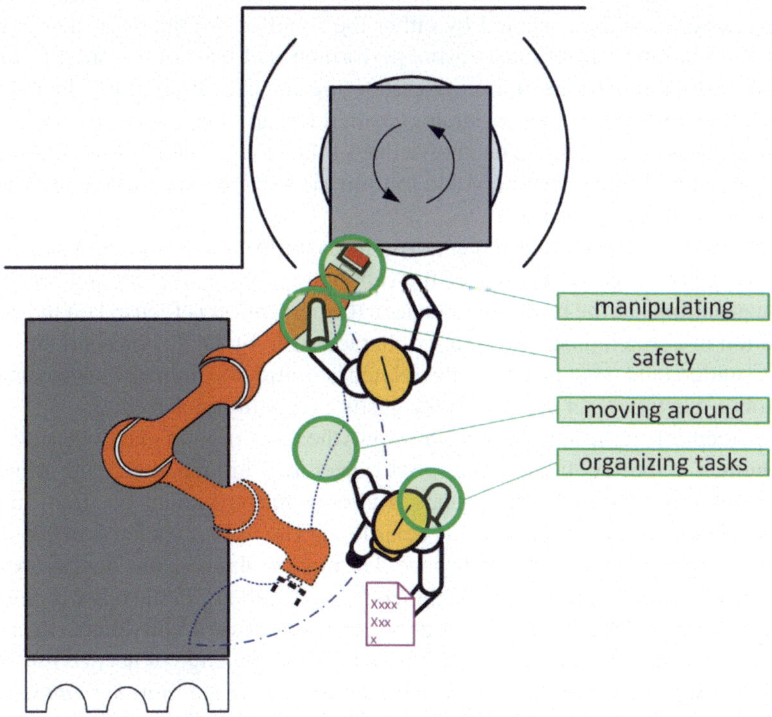

Fig. 4.6 Major enablers in PIROS experience: *manipulating* includes all interaction control functions for assembly tasks and robot actions on materials; *safety* includes risk assessment and reduction for hazards typically featured in human-collaboration (e.g., accidental contacts); *moving around* includes all motion planning functions for optimizing the movements of the robot system inside the shared workspace, in terms of both safety and productivity; *task planning* includes the software for assigning roles and activities to both human and robot agents

4.2.3 Proposed Architecture

The proposed approach to reach the required value of adaptability is to design modular solutions able to perform the sub-tasks. The task planner deals with the suitable scheduling of the sub-tasks. Operators can plan and/or compose these actions at runtime, considering user preferences and execution time in case of allocation of actions to robots. The major enablers of the proposed approach is highlighted in Fig. 4.6.

The actions have to be implemented in a modular fashion, allowing easy assignment swaps between human and robot. Moreover, the programming time should be compressed as much as possible, indeed the modules could be rescheduled on-the-fly depending on the current layout of the cell. The task planning description is reported in Sect. 4.3.3.

Many actions can be executed by either the robot or the operator, therefore the robot manipulation abilities must ensure performance at least of the order of magnitude of the operator in terms of force overshooting and execution time. The proposed approach is an impedance control strategy with reference shaping coupled with environment stiffness estimation. The knowledge of the environment allows the shaping algorithm to avoid force overshoot and to limit the needed execution time. Manipulation is described in Sect. 4.3.1.

In human-robot collaboration scenarios, the usability and the operator satisfaction strongly depend on the capability of the robotic system of reading and reacting to the human actions. In the proposed scenario, the perception is focused on the gesture recognition in a true industrial scenario, characterized by occlusions and noise. The goal is to understand what piece and the related activities the human wants to delegate to the robot. Section 4.3.4 deals with the gesture recognition activities.

The perception of the human position allows the robot motion planning to be aware of the human activities. Indeed, the robot trajectories influence and are influenced by the human activities. In the proposed approach, the human position during a task is modelled as a probabilistic 3d scalar field. The field associates an occupancy probability to every point in the workspace. The motion planner considers as obstacles the volumes where the probability is higher than a threshold. In the other regions, the algorithm computed the expected execution time and the variability due to the human presence. This allows the scheduler to have a better knowledge of needed timing and to act accordingly. Section 4.3.2 deals with the human-aware motion planning.

The continuous reconfiguration of the cell increases substantially the efforts to achieve a risk assessment. In fact, the associated risk assessment needs then to consider a much higher number of assumptions at design time. Planning and scheduling generate task allocation hard to list entirely at design time. With the layout and hardware changeability, the number of risk variables is increased by partially-known objects inside the shared workspace. Operators might undergo to errors or out-of-training behaviors, and may entail different mistaken workflows, such as circumventing instructions or attempting hazardous troubleshooting.

The proposed solution to overcome this challenge is a model-based Computer-Aided Risk Assessment. It supports the designer of safety solutions (together with the robot system manufacturer and other stakeholders) during the phases of hazard identification, risk estimation, and risk evaluation. The approach is dedicated to modeling the collaborative application with a formal model that can be checked and verified for a given objective property, which in case of HRC situations corresponds to the condition of negligibility of all risks associated with hazards potentially occurring during the task execution. Computer-Aided Risk Assessment is presented in Sect. 4.3.5.

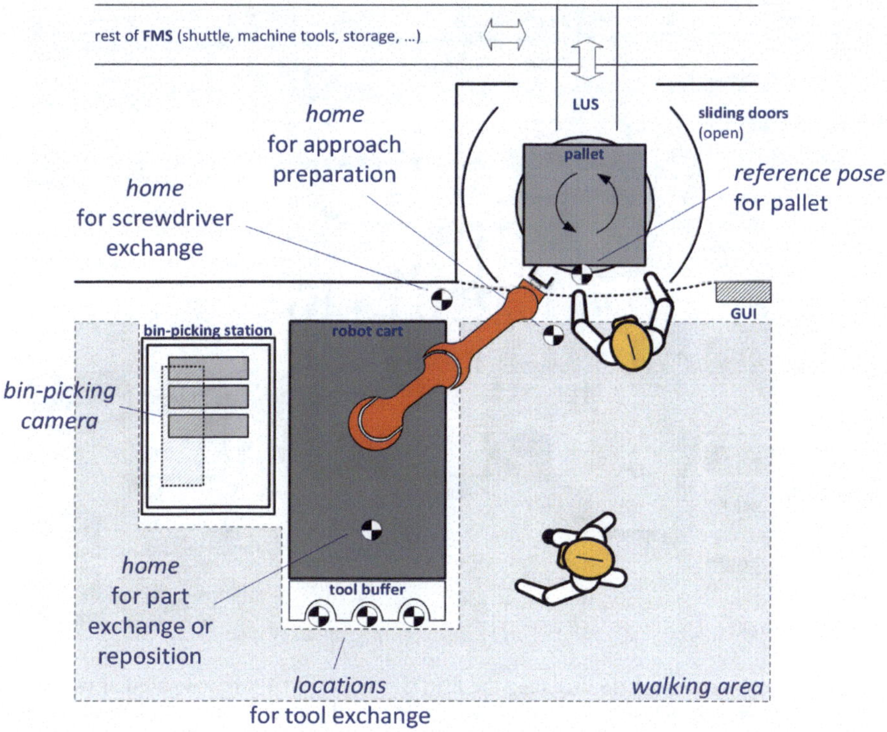

Fig. 4.7 Schematic top view of LUS layout, as depicted also in Fig. 4.8. *walking area* is the zone where operators are expected to move as part of their intended use of the system. The workspace of human operators (i.e., including the outreach of arms and hands) extends beyond the walking area, e.g., to the pallet *reference pose* and fixtures, to the *home for screwdriver*, etc)

4.2.4 Setup Description

The robotic cell is composed by a LUS, the robot equipped with the required tools and sensors and the human-robot shared workspace. Robot, tools and sensors communicate with a Ubuntu 16.04 laptop with ROS-Kinetic framework, this computer is integrated with the FMS by means of IEC61499 standard as described in Sect. 4.3.3.

The robot system is assembled into a load/unload station located in Cembre SpA. The layout of a LUS features a frontal access zone to the processing area, which is about 900 mm wide and is enclosed within sliding doors that are opened only when a pallet is ready for manual operations at the station.

The robot system (see Figs. 4.8 and 4.9) is composed of

- a lightweight robot KUKA iiwa R820, with 14 kg of payload;
- a Schunk WSG50 Co-Act gripper, custom-design pneumatic gripper for heavy load (based on Schunk PGN+ 160-2);
- a custom vacuum gripper;

Fig. 4.8 Frontal view of a LUS including the PIROS setup. The LUS is composed of two pallet units, of which only the left side one has been equipped with the robot system. Right hand side unit continues to operate manually (pictured with its sliding doors closed). The robot system is relocated on the left-end side of the left pallet unit (pictured with open doors and a pallet available in the processing area). The LUS, approx. 2300 × 3100 mm wide and high, is surrounded by the storage unit, where pallet are shelved and shuttled within the FMS. The portion of the FMS not visible in the picture include the full-length storage (approx. 30 m), another manual LUS, the shuttle delivery system travelling behind the LUSs, and 3 CNC machine tools (partially visible in the background, on the opposite side with respect the shuttle track). Notable elements of the system are labelled (Courtesy of Cembre SpA, Brescia, Italy)

- an electronic-controlled screwdriver. It enables the shared use of the tool by the robot and a human operator. The high required torque (up to 90 Nm) is balanced by a reaction carbon-fiber telescopic arm, directly attached to the tool. The tool is gravity-compensated and attached by the robot with a standard tool changer;
- Ensenso depth sensor deals with human tracking perception;
- A stereo-camera on-board system for object recognition in bin picking activities;
- Protective shells have been added around on-board cameras and the tool changer for collaborative usage purpose.

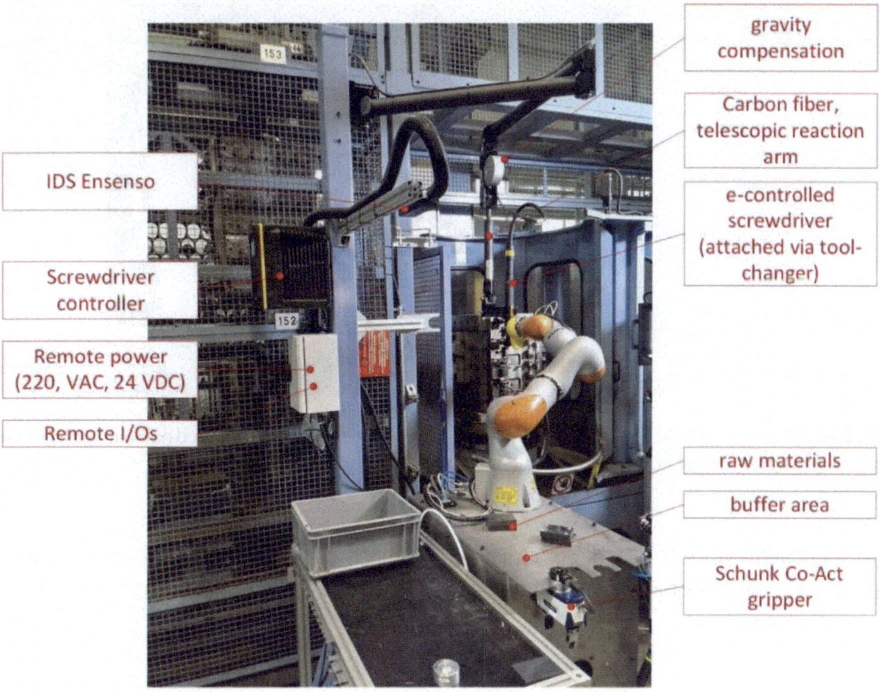

Fig. 4.9 Lateral view of the LUS depicted in Fig. 4.8, showing and listing several tools included into the robot system. The screwdriver storage point (see also Fig. 4.7) is visible at the center of the image, while the screwdriver is attached to the robot as the end effector. The buffer area on top of the manipulator cart is used for occasional re-positioning of parts or raw materials (Courtesy of Cembre SpA, Brescia, Italy)

4.3 Activities and Results

In this section the activities described in Sect. 4.2.3 will be described in details.

4.3.1 Manipulation

Within the application scenario, the requirements are mainly two:

- in terms of processing performances, the main quality metrics is associated with the limitation of force overshoots.
- Fast execution cannot be longer than manual insertion time (2–3 s) in order to maintain the scalability of the robotic solution alongside human tasks (see Fig. 4.10).[1]

[1]As an example, some fixtures experienced during tests and prototypes, featured 24 slots for processed materials, half of which requiring full flipping of metal parts before insertion.

Fig. 4.10 Example of CNC pallet featuring twelve 2-slot jigs: each part is flipped between individual slots and inserted in pegs (pictured during development stages)

and their achievement has been addressed through the design of a proper adaptive impedance-based controller, that has been optimized in term of minimization of the force tracking error.

4.3.1.1 Methodology

The controller design is based on a two-layer architecture, with a first layer wrapping a standard compliant robot control. Second, a layer for converting a target force into an impedance set-point is designed for estimating and adapting the overall robot-environment impedance. Specifically, the main goal of a stable force tracking control problem is related to limiting or eliminating any force overshoot. The problem is therefore formulated taking into account the coupled dynamics (controlled robot—interacting environment) at the established contact conditions with the pallet fixtures.

A key point in the controller design has been the online estimation of the environment stiffness, needed in order to proper calculate the control gains to achieve the target interaction dynamics. With reference to the general control scheme in Fig. 4.11, an Extended Kalman Filter (EKF) has been implemented in order to estimate properly environment stiffness to the reference shaping algorithm (more details and simulation/experimental validation are in [6]).

The impedance control loop design has been therefore designed in order to guarantee the achieving of a pure decoupled second-order impedance behavior for the controlled robot up to a reasonable frequency, of around 5 Hz.[2] Therefore, the target dynamics for the controlled robot should result in:

[2]The robot used, KUKA iiwa already displayed a decoupled behavior in the Cartesian space. However the approach can be easily extended also to other robots, by properly designing a control loop around the standard position controller for many lightweight industrial robots based on [7].

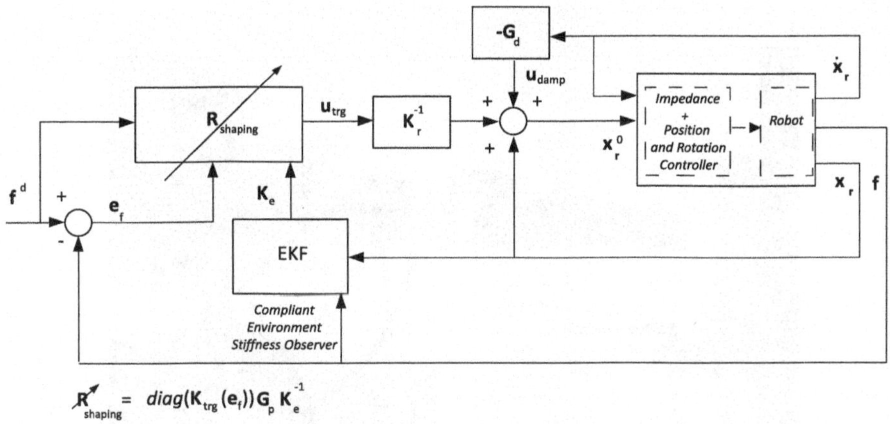

Fig. 4.11 Set-point deformation, including the environment observer (EKF)

$$\mathbf{M}_r\ddot{\mathbf{x}}_r + \mathbf{D}_r\dot{\mathbf{x}}_r + \mathbf{K}_r\boldsymbol{\Delta}\mathbf{x}_r = \mathbf{f}_r, \quad \text{with} \quad \Delta\mathbf{x}_r := \mathbf{x}_r - \mathbf{x}_r^0 \tag{4.1}$$

where $\mathbf{x}_r^0$ and $\mathbf{x}_r$ are the desired and actual robot positions respectively, and $\mathbf{f}_r$ is the external interacting force/torque (Fig. 4.11). In addition, the Cartesian stiffness $\mathbf{K}_r$, damping $\mathbf{D}_r$ and mass $\mathbf{M}_r$ have to present negligible extra-diagonal coupling terms with a good approximation up to few Hz. In particular, $\mathbf{D}_r = 2\,\mathbf{h}\,\mathbf{M}_r\,\omega_0$, where $\mathbf{h}$ is diagonal matrix of the imposed damping ratio and $\omega_0 = \sqrt{\mathbf{M}_r^{-1}\mathbf{K}_r}$ is the system pulsation.

4.3.1.2 Results

Developed controllers were able to optimize a hard-contact assembly process with tight mounting tolerances (e.g., H7/h6, as in Fig. 4.12) both in terms of force-tracking and execution time. The force overshoot f^{max} is computed at the highest peak magnitude w.r.t the target force f^d, the steady state force f^{ss} is averaged over 100 samples from the force overshoot f^{max}, the execution time t^{exe} is computed from the first contact until the opening of the gripper (end of the task) and compared with the expected target execution time t^d.

Table 4.1 shows obtained results for f^{max}, f^d and t^{exe} for $i = 1, \ldots, 5$ iterations using the setup in Fig. 4.12, while Fig. 4.13 shows the measured force during the assembly task during one iteration.

The experiment reports an average 9% overshoot and 1% error force and an average execution time of 1.13 s, corresponding to an improvement of 89% w.r.t. the target t^d. The control variable with KUKA iiwa is the position reference, making the force tracking in impedance mode being very accurate. The first peak and settle phase in Fig. 4.13 corresponds to the preengagement of the pegs, with the highest peak on reference being the actual forced insertion.

Fig. 4.12 Example configuration of a fixture assembly task (H7/h6 tolerance)

Table 4.1 Target force f^d, force overshoot f^{max}, steady state force f^{ss}, execution time t^{exe} are shown for each *i*th assembly. Experiments executed with the KUKA iiwa robot

Assembly results using the KUKA iiwa					
Assembly	1	2	3	4	5
f_i^d $[N]$	40	40	40	40	40
f_i^{max} $[N]$	43.3632	43.4049	43.5111	43.6693	43.6454
f_i^{ss} $[N]$	39.7509	39.8195	39.7514	39.6848	39.7678
t_i^d $[s]$	15	15	15	15	15
t_i^{exe} $[s]$	1.13	1.13	1.12	1.13	1.12

As an overall result, the avoidance of the force overshoots and a drastic reduction of the execution time of the assembly process proved the control strategy as an effective option for the collaborative pallet assembly scenario.

4.3.2 *Motion Planning*

In collaborative applications, human's and robot behaviours are coupled, and robot trajectory influences ergonomics and the way in which the worker moves affects the robot trajectory [8, 9]. The primary reason for altering a robot trajectory is safety, i.e., it may be necessary to stop or to slow down in order to avoid collisions. However motion re-planning may also be involved in re-tasking robot actions on-the-fly. This coupling has a relevant impact on task planning and scheduling activities in the state-of-the-art, however, the task planner/scheduler is not able to *reason* on the basis of uncertain robot execution times and changing human behaviours [10]. In

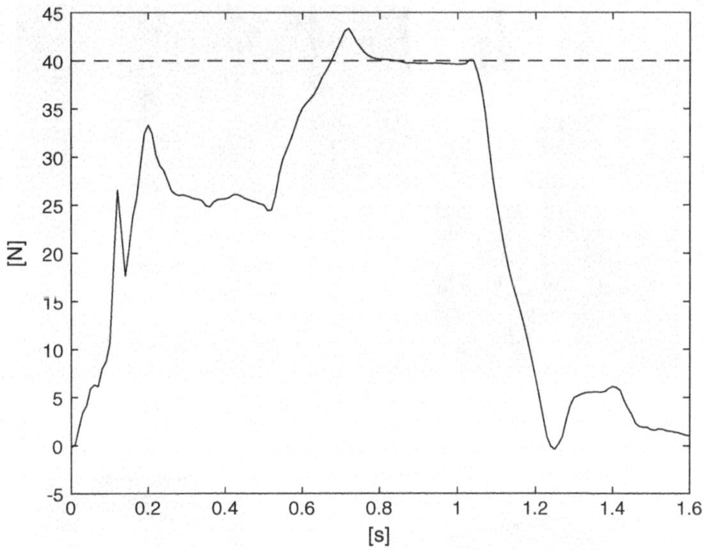

Fig. 4.13 Measured force versus target force during the experiment with KUKA iiwa

fact, currently available approaches for task planning are limited in reasoning about robot trajectories at spatial level [11, 12].

To overcome such limitation, within EuRoC, Team PIROS investigated motion planning where the timing of HRC tasks has been previously estimated for enabling the optimal planning and scheduling of e.g., assembly procedures [13].

4.3.2.1 Methodology

Within PIROS, the methodology consisted of a probabilistic time-description of the human occupancy close to the robot, and the estimation of how the overlapping of the robot trajectories with the probabilistic occupancy volume corresponding to the operator influences the trajectory time.

Specifically, a workspace segmentation has been performed considering the volume occupied by the human and the robot during their motion. Then, this segmentation is exploited for the definition of a set of Markov chains modelling human-robot interaction and allowing the estimation of the robot execution time. A complete overview of the methodology are in [14].

As show in Fig. 4.14, first, the Occupancy Grid representing the collaborative workspace is defined with a step of few millimeters, typically 5 mm. This resolution is sufficient to track properly the human movements as needed by the method, and it is coherent with the resolution of the Kinect One sensor. The two Kinect One sensors are placed so that occlusions are avoided, i.e., at least one Kinect One is always able to track human movements. The occupancy volumes are defined of the integral time

(a) Human Task

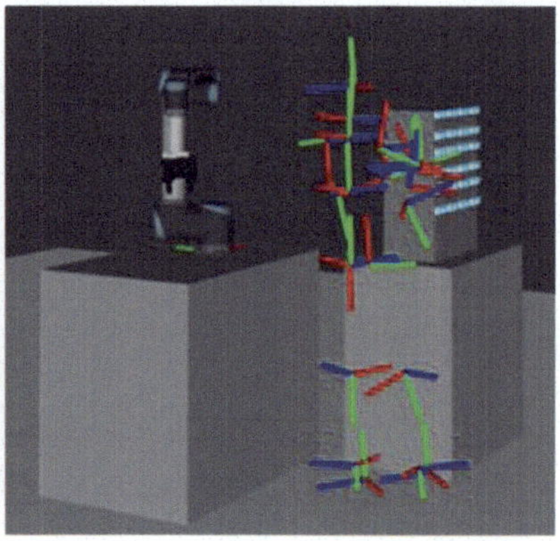

(b) Skeleton Acquisition

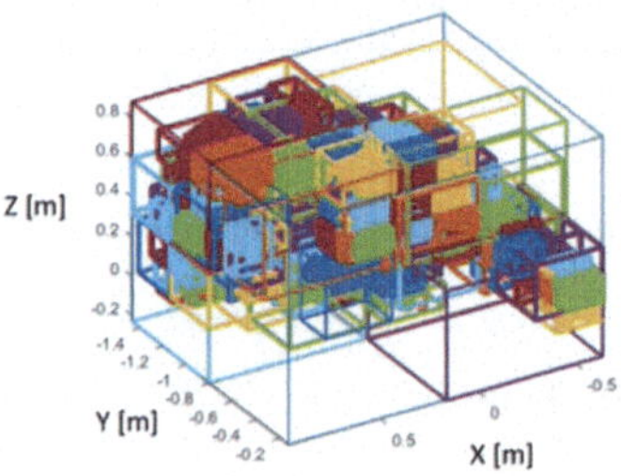

(c) Ful octree of the occupancy volumes givenby tracking the human skleton

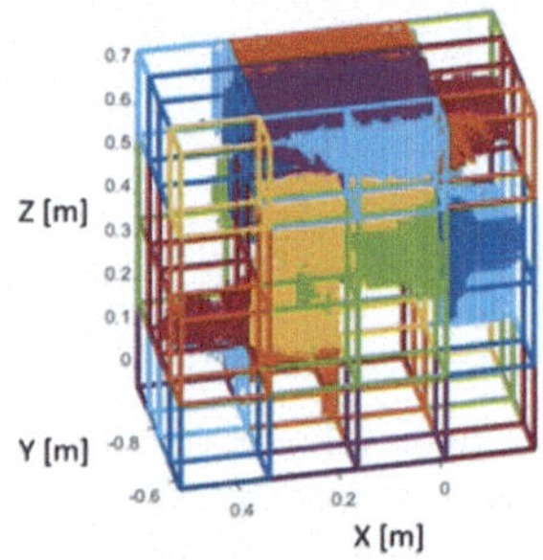

(d) Octree leaf bins

Fig. 4.14 The volume occupied by the human and the robot during the execution of their tasks is represented by two cloud points, than merged and organized into an octree. The leaf bins of the octree occupied by both the human and the robot will be used as nodes in a set of Markov chains

when at least a segment of the human is in the grid cell divide by the acquisition time. The computational complexity for the problem is reduced by means of a Octree algorithm. Given the task represented as an occupancy volume, it is possible to model a multi-order Markov Chain that can be used to model the interaction of the human with the robot, while the robot intersects the human workspace. Finally, through such multi-order Markov Chain is therefore possible to estimate the probability of collision, and the best estimation for the mean time of the robot trajectory when it is repeated many times close to the human.

Within EuRoC, the approach has been tested under multiple conditions. First, the approach was tested using a high number of simulated experiments. Specifically, a simulator of the real behavior of the robot was created and used to obtain an estimation of the robot expected time for randomly generated trajectories. The results coming from the simulation and from the proposed approach were compared in terms of mean execution time and standard deviation. Second, a set of experiments was conducted

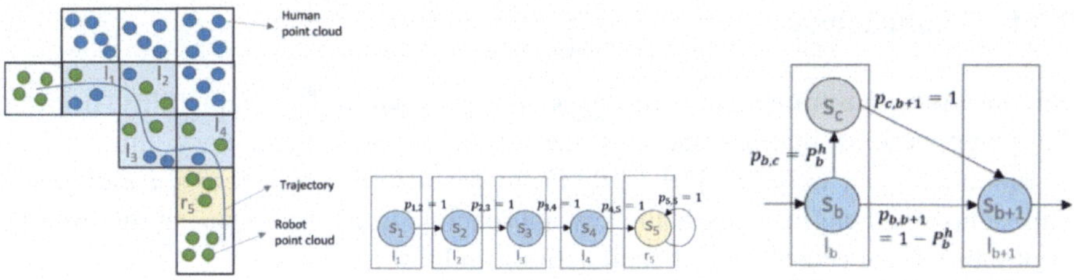

(a) Leaf bins of the octree and a trajectory that crosses the human workspace

(b) Octree with only leaf bins and a zero-order Markov chain built on it.

(c) Changes to be applied to the states and the transition probabilities of a Markov chain with order > 1 that models the stop of the robot.

Fig. 4.15 The volume occupied by the human and the robot during the execution of their tasks is represented by two cloud points, than merged and organized into an octree. The leaf bins of the octree occupied by both the human and the robot will be used as nodes in a set of Markov chains

on a real setup and compared in terms of obtained mean execution time with the results coming from the approach (Fig. 4.15).

4.3.2.2 Results

The results show to have relevant repercussions on the company profitability and productivity in terms of improved efficiency and reduction of idle time. Roughly, the method is able to estimate the mean time for a cooperative task with an error less than 5%. Given such accuracy in the mean time estimation, the software module developed in PIROS is able to select the best and mean-fastest trajectory among a wide set of many different paths.

4.3.3 Task Flow Management

Pallet assembly/disassembly displays some common features of highly dependable human-robot interaction tasks. The modeling and programming background is framed in the modular component-based software engineering. From a workflow standpoint, state-based execution is more effective than time execution: collaborative applications are loosely based on time sequences, although robot tasks in production are almost always programmed with deterministic timing because of simplicity and design attitude. In interactive tasks, however, it is recommended that the timing is not rigidly imposed by the robot for usability and comfort purposes [15, 16]. Instead, execution states may be triggered by events made by either robots or operators (see [17, 18]).

4.3.3.1 Methodology

Robot task programming has proved to be a major enabler for adaptability, because robotic applications analysis can be done on the basis of elementary units of logical functions (i.e., *actions*) [19, 20]. From a software engineering stand point, component-based engineering is the key approach [21, 22] for ensuring modularity, composability and reusability of such actions and tasks.

In PIROS approach such modular approach has been first pursued for a task programming standpoint, making use of the IEC 61499 standard [23, 24] for developing the application-level control structure, i.e., the top hierarchical layer for individual machines (e.g., robots). Composing tasks is a matter of assembling functional units (see Sect. 4.3.3.1), connected to the physical layer (i.e., device controllers). See an overview of the programming-deployment architecture in Fig. 4.16.

While this approach has been suitable in the initial stages of EuRoC challenge for local task programming by application developers or by end users through simplified graphical interfaces, for details see Sect. 4.3.3.1. The PIROS architecture later extended to include scheduling/planning problems. The allocation of human and robot tasks has been coordinated with the FMS shopfloor-level controller, extending the approach to a multiscale role and action allocation based on Flexible Timelines. An overview of an expanded programming-planning architecture are in Fig. 4.21 while details are in Sect. 4.3.3.1.

Task programming using IEC61499 function blocks

IEC 61499 defines architectures and standards for logical Function Blocks (FB) which are the natural modular and reusable entities constituting one of best know methodology in the spirit of component based engineering (see examples in [25]).

Robot control through IEC 61499 dates back from [26] to recent results [27, 28]. While exploiting the similar wrapping approach for encapsulating low-level features, our development is comparatively less centralized and provided a good degree of composability. All function blocks dedicated to operator's inputs and behaviour are more flexible with the possibility of being interrupted, with the possibility of introducing a broader spectrum of Temporal Allen's Relations [29].

The proposed approach consists of a layered (hierarchical) organization of FBs, associated to progressively low-level functionality:

- **scheduling level** that manages on-line the allocation and execution of applications based on internal/external requests. Requests can originate from operator decisions, external plant inputs, events in other applications which are running in parallel. Queuing requests and sorting the allocation of applications is one of the simplest and most effective ways to coordinate multi-applications workcells;
- **application level** that includes all the specific application instances, namely all the branches of interconnected tasks that reflect a desired work-flow or elementary sequences of tasks. The definition of applications as interconnected, orderly triggered sets of actions (encapsulated in FBs) is fundamental for a lean structure of this layer;

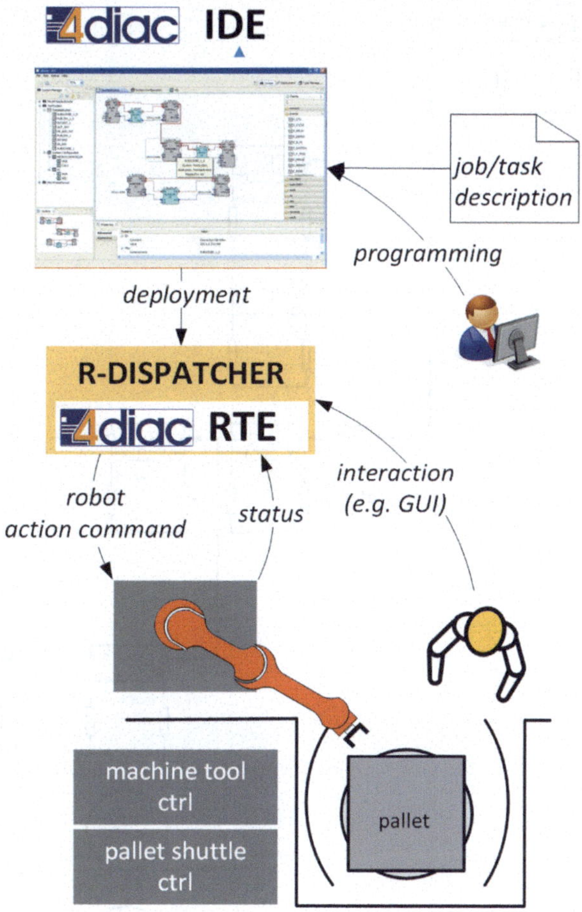

Fig. 4.16 Overview of the programming-deployment architecture

- **low level** (closed-loop control) which is dedicated to machine-level behavior. As far as closed-loop control and robot trajectory generation are concerned, the Robot Operating System framework (ROS) is one of the best candidates for being wrapped inside FBs.

More details are provided in [30]. The deployment of IEC 61499 solutions according to the approach in Fig. 4.17 is based on the composition of logic in the dedicated IDE (4DIAC) (Fig. 4.18).

Despite the "spaghetti chart" appearance, tasks composition in a IEC 61499 IDE is straightforward for e.g., PLC programmers. All relevant blocks for conditional execution and user-interaction are included (i.e., decisions, queries for options, wait for commands, alerts), although the user interface usability is limited.

Task planning with Flexible Timelines

Component-based engineering supported by IEC 61499 FBs is a powerful method for integrating local robot system scheduling and control, with factory-level supervisory

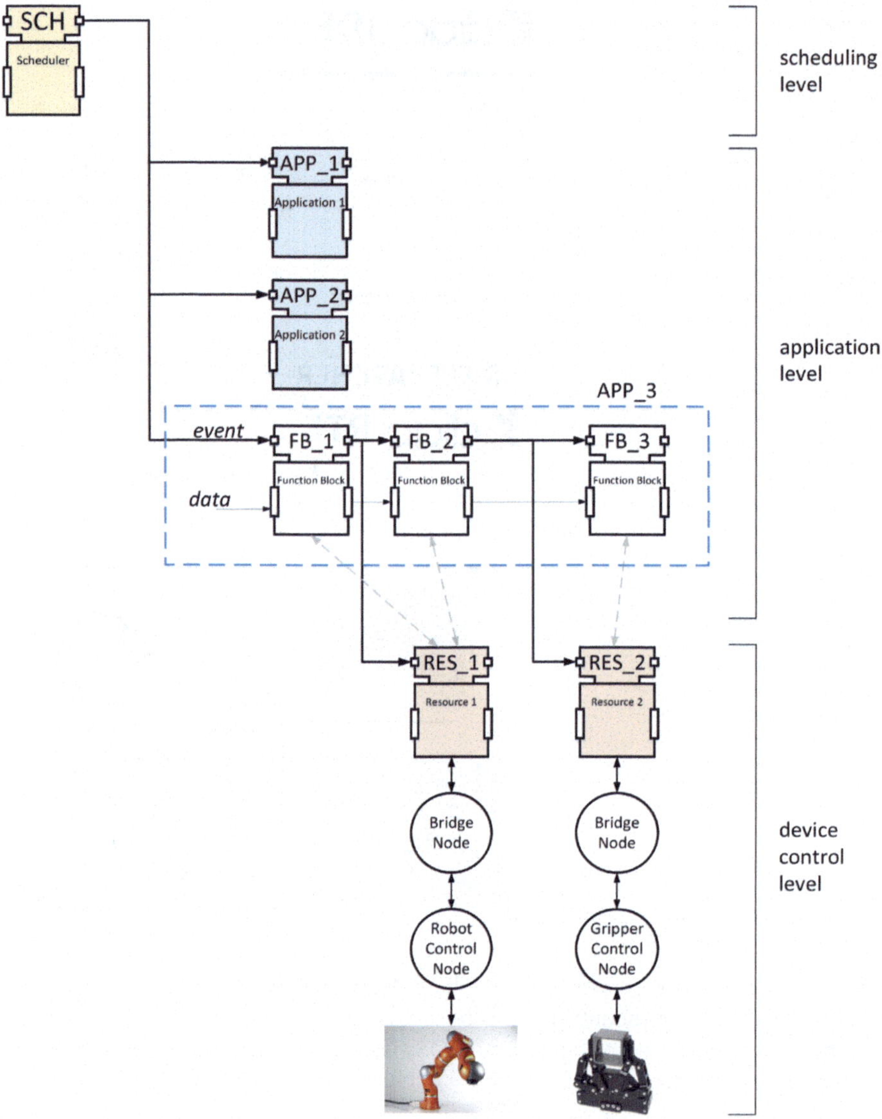

Fig. 4.17 Actions and function blocks

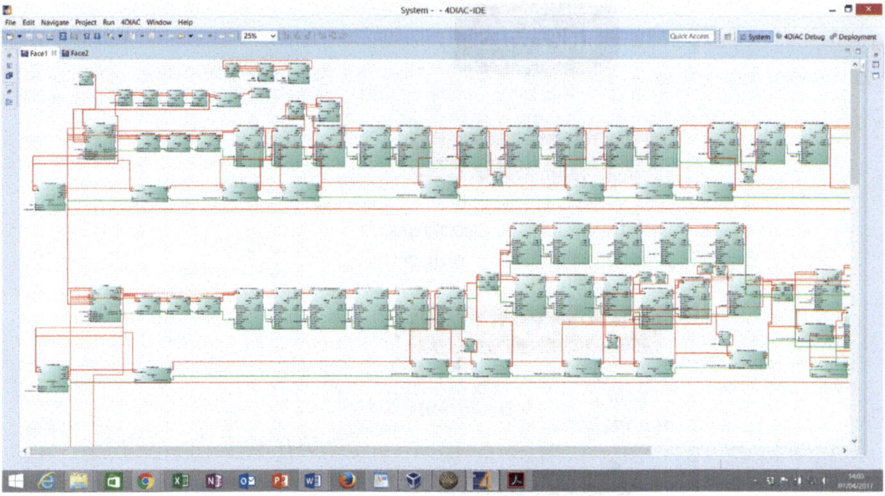

Fig. 4.18 Example of function blocks application in 4DIAC

systems, like the FMS supervision module developed by PIROS end user MCM S.P.A and called java Flexible Manufacturing eXecution (jFMX). The jFMX module is delivered in two versions: “level 1” for FMS control, “level 2” is interfaced with the factory ERP, increasing the scope and scheduling capabilities with respect to production objectives. The jFMX module is natively an even-driven architecture, whose functional logical units are CNC machine tool actions. While wrapping IEC 61499 FB to jFMX is straightforward, we found that the resulting hierarchy (from jFMX task scheduling to execution commands for robots) would have limited the flexibility of human-robot local role allocation.

While at the macro scale the jFMX module elaborates a list/queue of eligible tasks to be assigned to a LUS, the micro-scale R-PLANNER selects which single task to execute (i.e., calling for a specific element of the FMS queue), and how to assign the task actions between the robot system and the human operator (see Fig. 4.19).

The possibility of variously combining synchronous, simultaneous or asynchronously supportive types of interaction between humans and robots, allows the R-PLANNER modules a higher flexibility in task assignment phase but requires a detailed knowledge of the tasks at hand, and a full understanding of the risks involved with task-centric collaborations [31], e.g., drifts from coordination and deployed plans. The PIROS framework for robot planning is based on flexible timelines [32]. Inspirations and design motivations are derived from:

- long-standing approaches [33] that allow the robot to select and perform its tasks while taking into account explicitly the constraints imposed by the presence of humans, their needs and preferences;

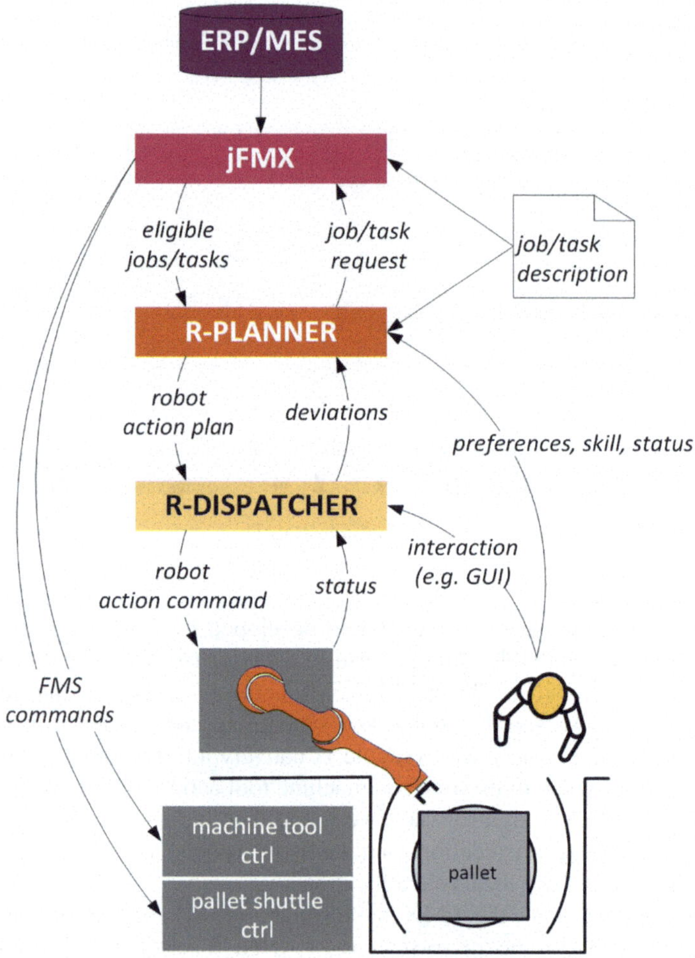

Fig. 4.19 Concept of vertical integration between MCM's factory-level control and robot control. (ERP = enterprise resource planning; MES = manufacturing execution system; jFMX = MCM machining supervisor; R-PLANNER = planner module of robot system; R-DISPATCHER = dispatcher module of robot system)

- and recent works [34] about flexible-timeline control architectures based on hierarchical task decomposition in which the hierarchy of the timelines reflects the aggregation of actions performed by the system resources (operator and robot).

Flexibility in task allocation or recombination between humans and robots (under the hypothesis that same tasks can be allocated to both agents), can be conveniently attained taking into consideration important aspect of human-robot collaboration, like operators' preferences (usability factors) and degree of strain (physical ergonomics factors). A generated plan (see R-PLANNER in Figs. 4.19 and 4.21) for assignment of specific actions may select different intensity and/or repetition patterns in pallet assembly, depending on the type of fixtures to handle, occurrences of different fixtures in the pallet queue, preference of operators for anticipating some specific tasks, etc.

4.3.3.2 Results

The vertical integration model in Fig. 4.19 is designed and developed making extensive use of ROS components and packages, in order to take advantage from asynchronous communication (publishing/subscribing) and sub-task execution. The basic logical unit remain the *action* (as in Sect. 4.3.3.1). The task and action model are encoded in form of NDDL files. The actual planning capability is provided by the planner component EUROPA, which encapsulate all the resources of the robot system in form of *reactors* (see Fig. 4.20). Reactors can be designed in functional hierarchies with senior components setting goals for junior elements, while receiving observations (state of part of the system in form of data recorded at runtime). Reactors have the capability of interfacing with executable (e.g., control or IO) nodes through the publishing/subscribing protocol.

On the basis of the EUROPA foundational technology for planning algorithms on timelines, the framework of task description, planner and reactors is collectively implemented by the T-REX ROS package (see Fig. 4.21).

Interfacing is necessary between T-REX task description and overall goals (e.g., pallet queue to be necessarily worked) and the resources hardware layer (e.g., robot controllers) (Fig. 4.22).

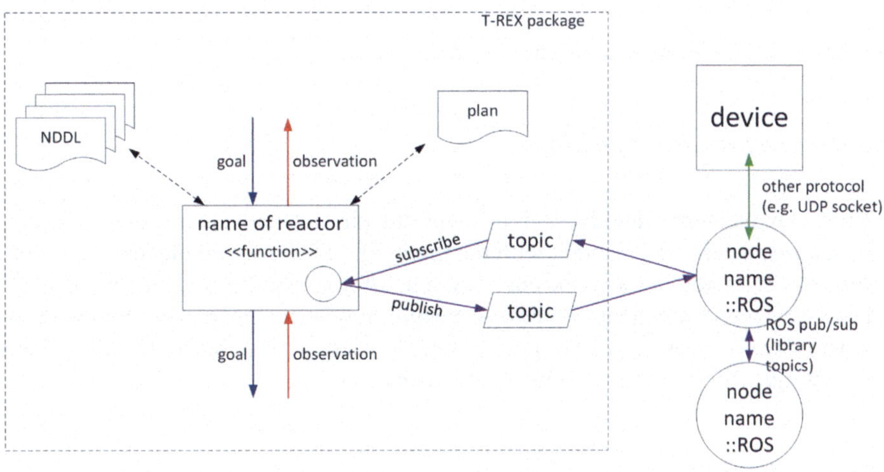

Fig. 4.20 Integration between T-REX package and ROS framework

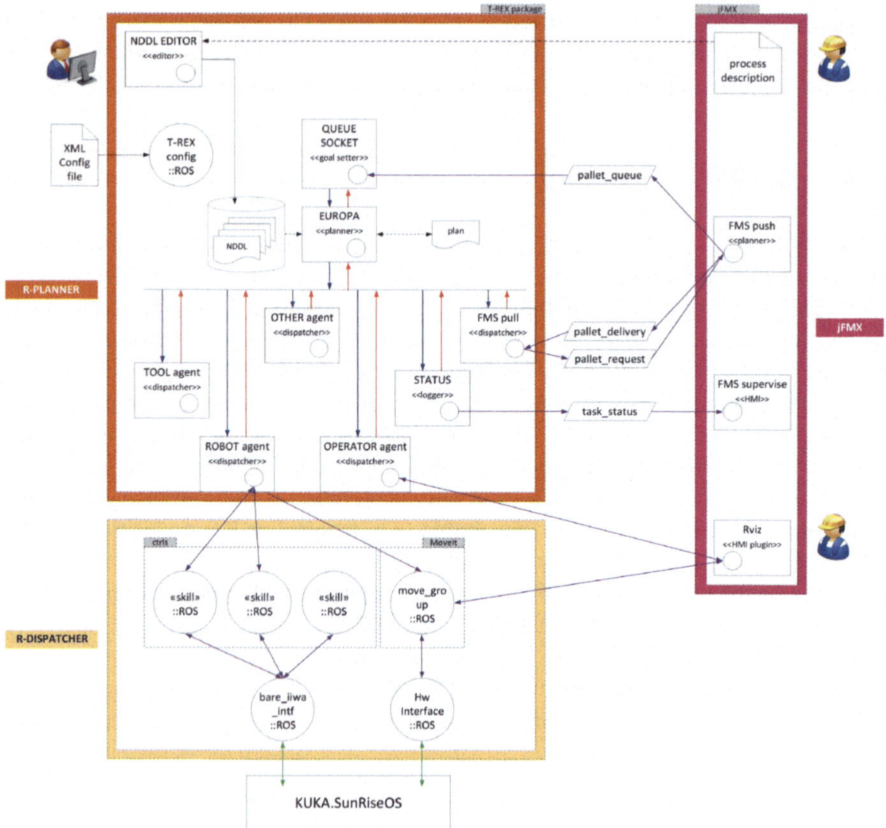

Fig. 4.21 T-REX ROS implementation of the framework

4.3.4 Gesture Recognition

Gestures are meaningful body motions involving physical movements of the fingers, hands, arms, head, face, or body with the intent of: (1) conveying meaningful information or (2) interacting with the environment. They constitute one interesting small subspace of possible human motion. A gesture may also be perceived by the environment as a compression technique for the information to be transmitted elsewhere and subsequently reconstructed by the receiver.

4.3.4.1 Methodology

Gesture recognition is commonly supported by 3D cameras (stereo vision), sensors for Point Cloud (e.g., LEAP motion), 2D color cameras. Other wearable approaches, based on e.g. accelerometers and gyroscopes are also used for very specific applica-

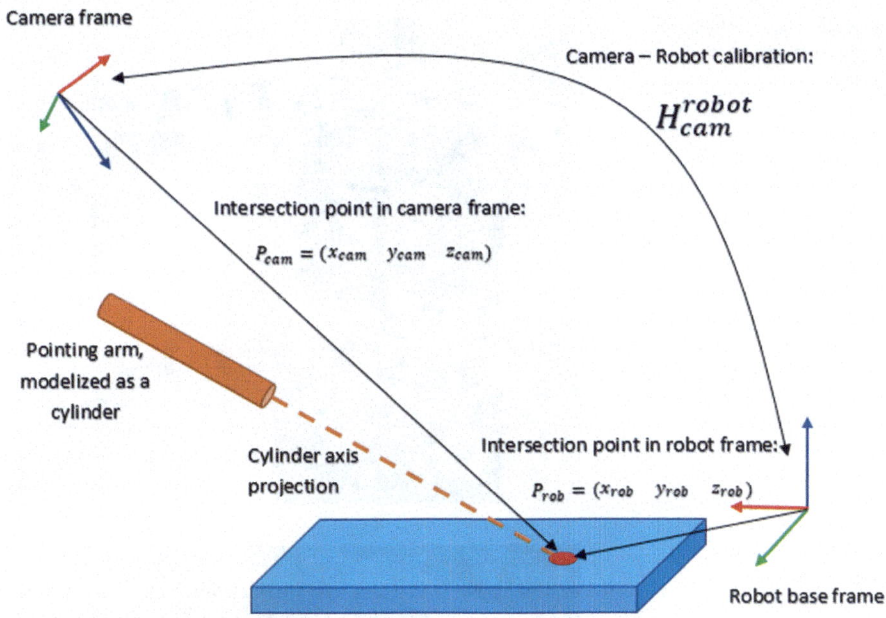

Fig. 4.22 User pointing to an insertion cell on face 2

tions questionably applicable in many manipulation-intensive tasks. Gesture information is derived by e.g. depth segmentation and skeleton tracking with 3D sensing, or by color segmentation (detection of hands, head or tags) with 2D.

The industrial challenge investigated within EuRoC have been to catch rapid human detection in cluttered environments, since such task has high requirements in terms of gesture recognition in variable conditions, with efficient processing of point clouds. Furthermore, the reference setup is very prone to occlusions and environmental noise. Specifically the effort have been dedicated to perform a filtering operation to detect stable gestures, i.e., the target location on tombstone is provided by the operator with gestures (e.g. finger-pointing an empty slot onto the tombstone). Specifically, a gesture is considered stable when a number N of consecutive detected points lie within a neighborhood area of radius R. N and R are configurable parameters in the pointing gesture algorithm (see Fig. 4.23).

The pallet is defined as a grid structure. By analyzing the coordinates of the intersection point obtained in the pallet surface frame, it is possible to return the identifier of the cell the worker is pointing at. The design is intended to be modular w.r.t. the various appearances of fixtures that are possible attached to a pallet (Fig. 4.24).

Fig. 4.23 Basic schema of the pointing gesture algorithm

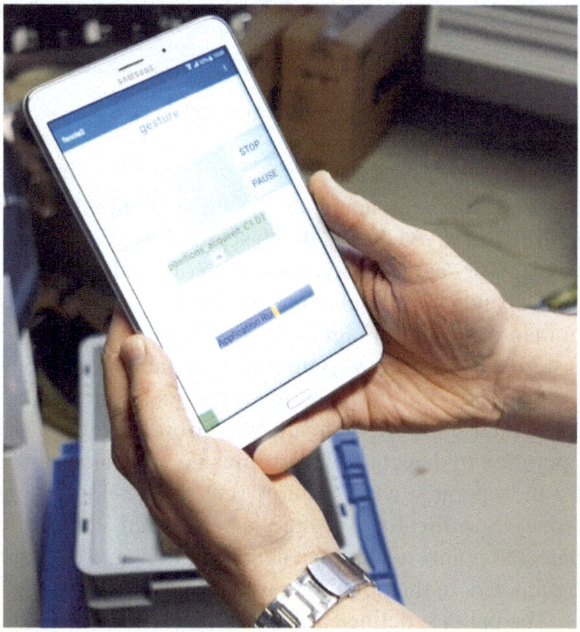

Fig. 4.24 Recorded cells inside the app

4.3.4.2 Results

The experiments in the freestyle demonstrated the effectiveness of the methodology, with a mean time needed for the detection of the gesture of less than 0.5 s, and with a percentage of success of greater than 90%.

Furthermore, during the showcase session, the industrial user considered the methodology robust and comfortable, and the usability was ranked as high.

4.3.5 Safety

Variability in task assignment between humans and robots, in addition to the high rate of collaboration by humans, increases the range of interaction situations possibly involving hazards. In fact, the close proximity and frequent interactions between robots and humans expand the *intended uses* (i.e., behaviors and tasks that do not deviate from the designed/programmed plan) of the collaborative robotics system. The associated risk assessment need then to consider a much higher number of assumptions at design time. Additionally, planning modules (see Sect. 4.3.3) generate task allocation which might be hard to list and evaluate entirely at design time. Considering, finally, the layout and hardware changeability (to adapt to different workflows or materials), the number of risk variables is further increased by partially-known objects inside a shared workspace. On top of the intended use *variability* (see Fig. 4.25), a standard, i.e., manual, risk assessment is challenged also by the *complexity* of situations introduced in a collaborative application, including, but not limited to, open ended environment whenever mobile solutions are involved and trajectories are not known in advance, and variable agents assigned to the application. This latter situation is typical of dynamic assignment of personnel or different

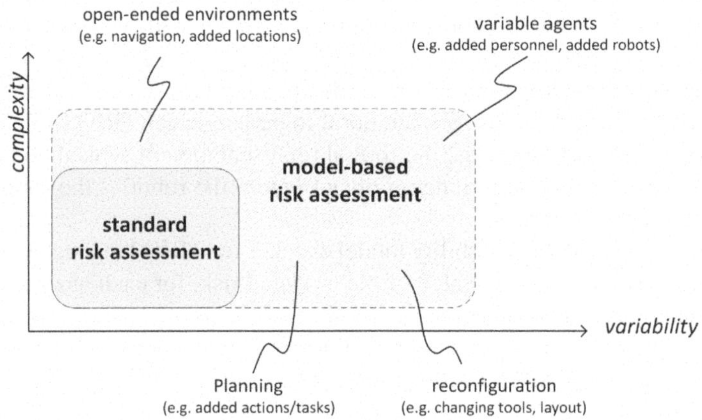

Fig. 4.25 Risk assessment, variability versus complexity

hardware resources to a robotic station, as it might be the case of a collaborative assembly task in a LUS.

Considering human aspects, operators might make errors or out-of-training behaviors, and may entail different *mistaken* workflows (e.g., circumventing instructions or attempting hazardous troubleshooting). The customization of applications, e.g., setting a preferred comfortable robot trajectory or altering the sub/task assignment during run-time execution (see Fig. 4.5), usually improves the usability of a system, but the hazard/risk distribution is more dependent on synchronization capability of operators, their overall situation awareness when supervising multiple tasks and ability to react/avoid hazardous situations during transient conditions (e.g., actions for catching up with a robot operation).

Hence, it is crucial that the design of the task and workspace provides suitable protection for operators in changing conditions, and that the actual deployment of applications conforms to such a safe design. PIROS experience provided an outstanding use case for introducing model-based, Computer-Aided Risk Assessment (CA-RA) with the purpose of supporting the designer of safety solutions, together with the robot system manufacturer and other stakeholders, during the phases of hazard identification, risk estimation, and risk evaluation. The approach is dedicated to modeling the collaborative application with a formal model that can be checked and verified for a given objective property, which in case of HRC situations corresponds to the condition of negligibility of all risks associated with hazards potentially occurring during the task execution.

4.3.5.1 Formal Modeling and Verification of Safety Conditions

The safety verification of target tasks is based on a formal model representing the elements of a collaborative application (operators, robots and environment), together with models of hazards (e.g., contacts), risks and risk reduction measures (e.g., limit energy). A formal verification tool is used to comprehensively explore and check all possible interactions between robots and human operators, with respect to *risks*, along the designed task. Remarkably, the verification procedure is in charge of exploring also the *unintended* interactions, where some potential hazards may occur.

The model is formulated using a temporal logic language (TRIO [35]) featuring a quantitative notion of time, e.g., the formal representation of typical human-robot interaction situations like "reaching an object *before* the robot", "the robot *always* stays in a given region", etc.

Zot [36] is a bounded satisfiability model checker for TRIO language [37], where the property to satisfy is a low value of the modeled risk, for each hazard identified during the automatic model checking phase. In case the property (i.e., low risk) is not satisfied, Zot provides a counterexample witnessing a system execution that violates the property (i.e., a residual hazardous situation).

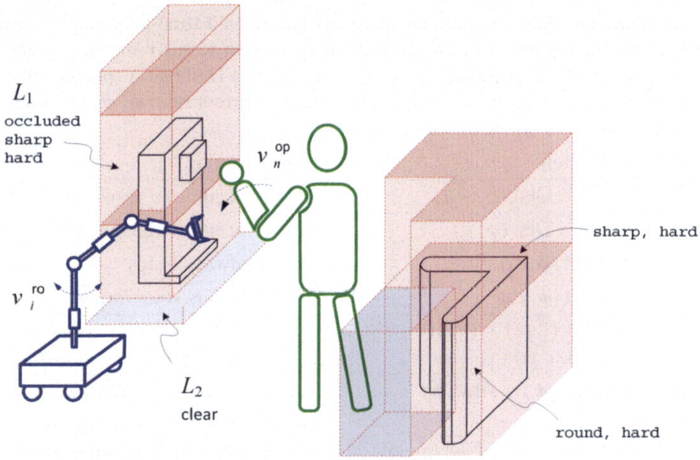

Fig. 4.26 Conceptual scheme of the robotic cell

TRIO formulae are designed and suited[3] for expressing the temporal relationships of human-robot collaboration situations, including the interactions with the environment (e.g., machinery, fixtures, etc), that could affect safety. The formal model, in fact, includes both agents (e.g., humans, robots) and those interactions which can be classified as hazards (e.g., collisions, entanglements, shears, etc). For example, an accidental contact between an operator and the robot system may occur with different hazard profiles depending on the timing of arrival in the same location, e.g., both human and robot converge to a single point are likely entangling in a mutual crush; a robot arm moving away from a pallet fixture while the operator is accessing is more likely to determine a dynamic impact).

With reference to [40, 41] for complete details, the model is formulated in terms of TRIO predicates. First, three main sets of elements are defined, one for each component of a collaborative application: the human body $\mathscr{O}$, the robot system $\mathscr{R}$, and the environment $\mathscr{L}$ with a given layout (see Fig. 4.26):

- the set $\mathscr{O}$ is a collection of eleven discrete body regions $\{O_{head}, \ldots, O_{leg}\}$ [42], where each region $O_i = \langle p_i, v_i, m_i, k_i \rangle$ is characterized by a set of predicates, such as for example p_i, representing the location of body region i, v_i, representing its velocity, etc.
- the set $\mathscr{R}$ includes elements of a robot, with predicates $R_j = \langle p_j, v_j, f_j, m_j, shape_j \rangle$ for each of its n links.

[3]We adopted TRIO formal language and its supporting tool Zot because of the generality of the language, which can be applied in natural way to different types of applications (together with MTL, TRIO is one the first temporal logic languages dealing with time in a metric way) and for the efficiency of the verification in comparison with the state of the art [38]. For a comprehensive survey and comparison among various logic languages tailored towards modeling time dependent phenomena see [39].

Table 4.2 List of derived TRIO operators. In particular, given a time-dependent formula ϕ (i.e., a term representing a mapping from the time domain to truth values) and a (arithmetic) term t indicating a time distance (either positive or negative), formula $\mathsf{Dist}(\phi, t)$ specifies that ϕ holds at a time instant at a distance of exactly t time units from the current one; ϕ, ψ denote propositions, and v is a variable and d is a constant value

TRIO operator	Definition	Meaning
$\mathsf{Past}(\phi, d)$	$d > 0 \wedge \mathsf{Dist}(\phi, -d)$	ϕ occurred d time units in the past
$\mathsf{Futr}(\phi, d)$	$d > 0 \wedge \mathsf{Dist}(\phi, d)$	ϕ occurs d time units in the future
$\mathsf{Alw}(\phi)$	$\forall t(\mathsf{Dist}(\phi, t))$	ϕ always holds
$\mathsf{Som}(\phi)$	$\exists t(\mathsf{Dist}(\phi, t))$	ϕ occurs sometimes
$\mathsf{Until}(\phi, \psi)$	$\exists t(\mathsf{Futr}(\psi, t) \wedge (\forall t'(0 < t' < t) \Rightarrow \mathsf{Dist}(\phi, t')))$	ψ will eventually occur and ϕ will hold till then
$\mathsf{Until_W}(\phi, \psi)$	$\mathsf{Until}(\phi, \psi) \vee \mathsf{Alw}(\phi)$	weak until: ψ may never occur in the future
$\mathsf{WithinF}(\phi, d)$	$\exists t(0 < t < d \wedge \mathsf{Dist}(\phi, t))$	ϕ will occur within d time units

- the model $\mathscr{L}$ of the layout includes a tuple of predicates $L_k = \langle shape_k, material_k, obst_k \rangle$ for each of the sections in which the layout is divided. Some predicates, like $shape$, $material$ or clearance/occupancy $obst_k$ carry safety-related information.

The model evolves through states, assigned according to predicates, e.g., $p_i^{\mathtt{ro}} = L_k$ means the presence of the i-th link of the robot inside the k-th region of the layout. The temporal logic is formulated in terms of first-order connectives, operators, and quantifiers, as well as some composed predicates (e.g., $\mathsf{InSameL}_{ijk}$, etc) based on a modal operator, called Dist, that relates the *current time* to another time instant (see list of operators in Table 4.2).

In this way, the formal model defines formulae that capture various safety-related aspects of HRC applications, notably including hazards. For instance, a hazardous *quasi-static* (*Qs*) [42] contact $\mathsf{hzd}_{ijk}^{\mathsf{Qs}}$ of constrained type (for example, crushing) that occurs in section L_k of the layout between the i-th robot link (for example, the end-effector) and the j-th body part (for example, the human hand) is formulated as:

$$\begin{aligned} \mathsf{hzd}_{ijk}^{\mathsf{Qs}} \Leftrightarrow\ & \mathsf{InSameL}_{ijk} \wedge \mathsf{moving}_{R_i} \wedge \mathsf{Past}\left(\mathsf{Sep}_{ij} > \mathtt{close}, 1\right) \\ & \wedge \mathsf{Past}\left(p_i^{\mathtt{ro}} = L_k \wedge p_j^{\mathtt{op}} \neq L_k, 1\right) \\ & \wedge (obst_k = \mathtt{occluded} \vee \exists R_m \in \mathscr{R}\,(\mathsf{InSameL}_{imk} \wedge i \neq m)) \end{aligned} \tag{4.2}$$

where $\mathsf{Past}(F, 1)$ means that F held in the previous instant from the current one, $\mathsf{InSameL}_{ijk}$ and moving_{R_i} are abbreviations for formulae describing, respectively, that R_i and O_j are in the same layout section, and that R_i is moving, and Sep_{ij} is a predicate capturing the current separation between R_i and O_j. The last two lines of the formula formalize, respectively, the situation of R_i and O_j getting closer to one another (with the robot arriving in L_k *before* the operator), and the presence of obstacles (i.e., the occlusion of section L_k) in that condition, or the internal crunching of arms inside a manipulator.

A hazard is identified as soon as the model state matches a condition defined as in 4.2 (recall that the actual state is generated at verification time, and not assigned *a priori*). Identified hazards are then estimated in terms of risk, whose score is assigned using TRIO predicates that capture the severity and occurrence of hazards according to normative safety methodologies [43]. In particular, the hybrid method for risk estimation in [44] is used for setting severity with discrete scores in $\{1, 2, 3, 4\}$ and occurrence factors (frequency of situations, probability of errors and failures, probability of the avoidability of the situation) with discrete values cumulated in an aggregate score in $\{1, \ldots, 15\}$. A discussion on the determinism of score assignment and update of values is provided in [45].

Upon estimation of unacceptable risk level, verification of the formal model highlights a condition of unsatisfiability of the target safety property. The model needs to be updated including Risk Reduction Measures (RRM), which are the models of safety strategies to be implemented. For instance, a collision avoidance method for preventing the contact hazard hzd_{ijk} compliant with the safety mode Speed and Separation Monitoring (SSM) [42] is forumlated as:

$$\begin{aligned}\mathsf{RRM}_{ijk}^{SSM} \Rightarrow & \left(\mathsf{Sep}_{ij} < \mathsf{Sep}_{min} \Rightarrow \mathsf{Futr}\left(v_{ij} = \mathtt{none}, 1\right)\right) \wedge \\ & \left(\mathsf{Sep}_{min} \leq \mathsf{Sep}_{ij} \leq \mathsf{Sep}_{mid} \Rightarrow \mathsf{Futr}\left(v_{ij} \leq \mathtt{mid}, 1\right)\right)\end{aligned} \tag{4.3}$$

where $\mathsf{Futr}(F, 1)$ means that F holds in the next instant from the current one, and Sep_{min} and Sep_{mid} are two thresholds on the distance between O_j and R_i, moving at relative speed v_{ij}. Sep_{mid} is used as a threshold for slowing down R_i, so to maintain a Sep_{min} threshold.

Recall that after the update of the model with RRMs, the hazard identification and risk estimation steps described above are repeated along the *new* evolution of states of the model, for checking the satisfiability of the safety property, which needs to be verified for each combination ijk identified as potentially critical:

$$\mathsf{Alw}\left(\begin{array}{l}\mathsf{risk}_{ijk} \leq \tau \vee \\ \mathsf{risk}_{ijk} > \tau \wedge \exists y \left(\begin{array}{l}\mathsf{RRM}_{ijk}^{y} \wedge \\ \mathsf{Futr}\left(\mathsf{risk}_{ijk} \leq \tau, 1\right)\end{array}\right)\end{array}\right) \tag{4.4}$$

where τ is the risk threshold value decided at design time, usually set to $\tau = 1$ (negligible/low risk) for scores $\mathsf{risk}_{ijk} \in \{0, 1, 2\}$ assigned using the ISO/TR 14121-2 hybrid method.

Such mechanisms of generation and evaluation of states of the model in every instant of its evolution are enabled by the Zot tool,[4] which implements several techniques—employing a so-called bounded satisfiability checking approach based

[4] Available from github.com/fm-polimi/zot.

on off-the-shelf Satisfiability Modulo Theories (SMT) solvers—for checking the satisfiability of TRIO formulae in an automated manner [38].

4.3.5.2 Experimental Results for PIROS Setup

The operational scenario and layout (see Fig. 4.2) are reproduced in the formal model (see Fig. 4.27). The robot is supposed to move between L_8 and the pallet repetitively, to bring and load pieces on the pallet. Each piece first be placed in L_4 and kept held until human operator terminates the part manipulation task.

Here we briefly report a couple of outputs and refer you to find more details on customizing the formal model in [45] and formalizing the executing task (i.e., UML profiles to logic formulae) [46]. In the experimental results reported in the rest of this section, only one potential feasible trace has been analyzed which about 410 sec long, verification bound is set equal to 40 time units, and each action takes at least three time units to complete.

First, the model is checked against the safety property without including RRM module to show that there are risky situations generated by the model that requires mitigation. Figure 4.28 pictures an example of occurrence of a hazard with a risk higher than two. There is no RRM to monitor and constraint the values of force and velocity. Hence, the severity of potential collisions could be quite high.

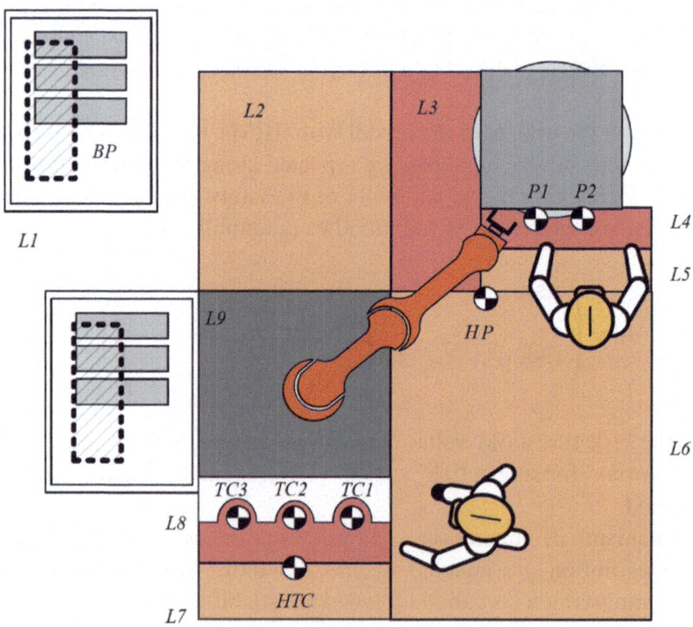

Fig. 4.27 Layout of the use case with its critical zones identifies with colors

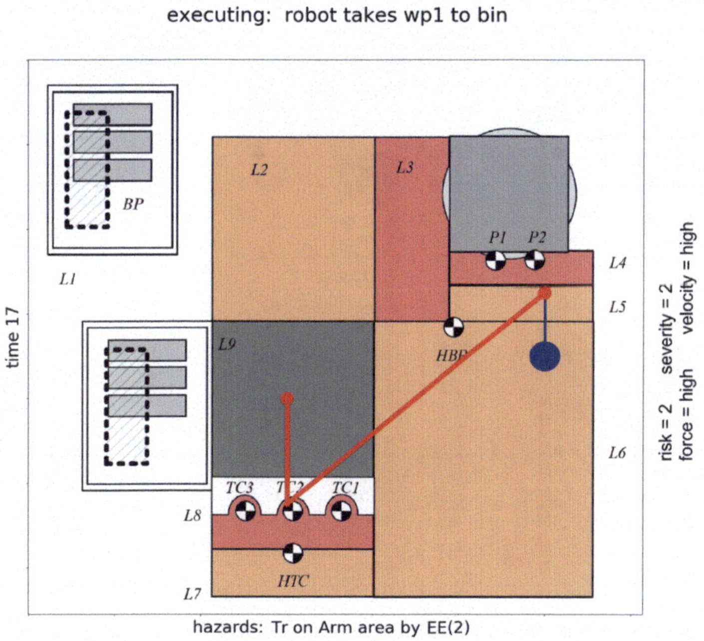

Fig. 4.28 This is an example of a situation in which RRMs are excluded from the model. Operator head and arm are depicted in blue, while robot links and endeffector are in red. The operator arm and endeffector are collided at $t = 17$ and the risk value exceeds the tolerable risk threshold

Second, we include RRMs in the model to show that same situations detected above are not critical as before because of RRMs. In this case the negation of safety property is never true and thus the tool reports **UNSAT**. However, in order to show how the system works, we removed the safety property and got a possible execution trace which is shown in Fig. 4.29. The safety of execution does not mean that hazards do not happen at all. The operator and robot have to be present in same locations to execute several actions and hazards are inevitable. However, RRMs eliminate the associated risk value to each hazard. For example, $t = 12$ in Fig. 4.29 is an instance of hazardous system state in-which the caused severity of the hazards is constrained by reducing the velocity of the moving robot parts.

The formal model is also able to capture hazardous situations that are caused due to human erroneous behavior. For example, Fig. 4.30 shows the same time unit in two different verification runs: one containing RRMs and the other one without RRMs. Both snapshots picture a time unit when the operator tends to execute an inspection at the wrong time. The operator should not execute an inspection while the robot is moving and getting close to the pallet, because it could cause a `Tr` hazards, as detected by the tool as well.

The critical difference between the time unit depicted in 4.30a and the one in 4.30b is the risk value. When RRMs are considered in the model, the risk value will not grow up to two since RRMs are invoked as risk reaches value one and try to

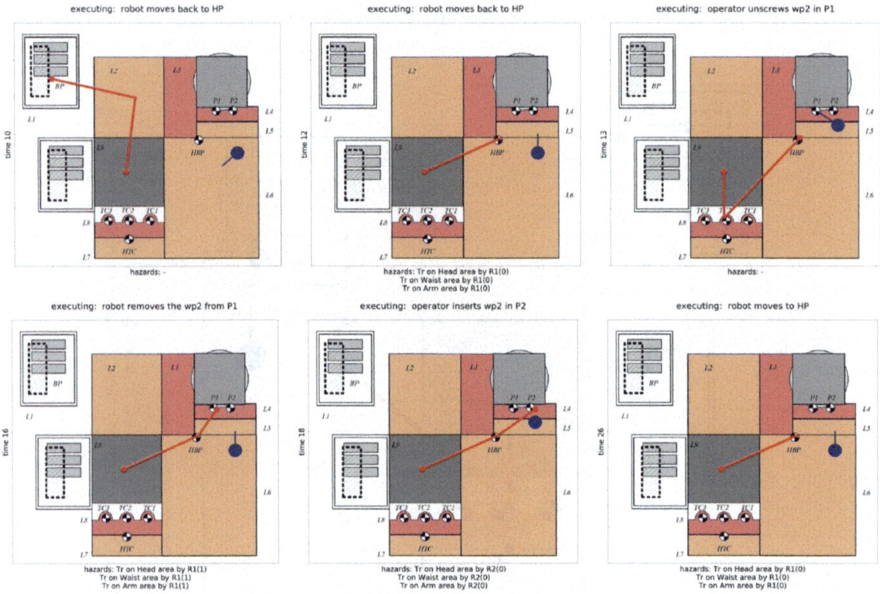

Fig. 4.29 An example of partial task execution in presence of RRMs which shows whenever hazards are detected, their risk is tolerable (less than two)

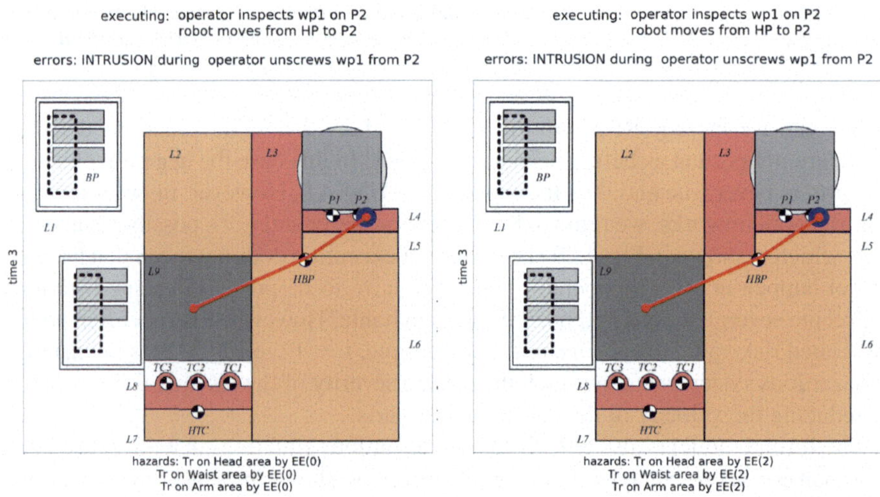

Fig. 4.30 An instance of an intrusion error

maintain or reduce the risk by reducing severity via constraining velocity and force values. The picture on left has two ongoing RRMs that keep the relative force less or equal than `low` (meaning: `mid` or `low`) and the relative velocity less or equal than `mid` (meaning: `mid`, `low` or `none`). The relative parameters here are pertaining to those parts of human body and robot parts which are involved in the definition of a detected hazard.

The model update is an explicit design action, which is ultimately under the responsibility of an overall safety supervisor. Different types of measures y (i.e., different RRM^{y}_{ijk}) can be chosen for mitigating risk_{ijk}, depending on the capabilities of the robot system (i.e., controls, sensors, etc.), and also on the possibility to modify the task itself, if necessary. As a consequence of the compliance with safety requirements (i.e., RRMs are necessary), production routines may be variously affected by the RRM strategy. For instance, slower safe modes in robot controls, re-arranged layouts, or forcing constrained action sequences by controlling access to locations, all have a substantial impact on the organization and/or efficiency of tasks.

4.3.6 Global Key Performance Index

Beyond the challenges related to each PIROS module, the overall approach has been investigated. Indeed, the PIROS challenge has been to deploy an ecosystem of modules that may improve dramatically the factory productivity and quality, not only tailored on the FMS use case, but with the ambition to be generally applicable to a wide field of applications.

The global Key Performance Index (KPI) for this objective is defined as the Added Value Index (AVI), which is the normalized ratio of high-ranking sub-tasks out of all sub-tasks:

$$AVI = \frac{\sum_i st_{i|r=3}}{\sum_i st_i} \tag{4.5}$$

where the rank r is a value-added score defined by the user, and st_i is a generic sub-task. Specifically,

- $st_{i|r=3}$ are sub-tasks of rank 3 like visual inspection, measurement on fixtures, resolution of faulty conditions on jigs or parts, etc.;
- lower value tasks—or sub-tasks that are labour-intensive—, e.g., pick and place, preparing insertion of parts in jigs, screwing/unscrewing, etc are ranked 1 or 2 by design since they introduce less value added.

When low-rank tasks are removed form the denominator in (4.5), the index increases, reflecting the fact that "better" sub-tasks are denser in operators' schedule (see an example in Table 4.3).

Simple, intuitive procedures for reviewing task/sub-task assignment before launching the task are fundamental for increasing AVI figures. The main indicator of performance is the seamless vertical integration of modular components able to share

Table 4.3 Added value index (AVI) index computation for a real case

Description	Rank	Default assignment	Optimal plan
Sub-task A	1	H	R
Sub-task B	3	H	H
Sub-task C	2	H	R
Sub-task D	3	H	H
Sub-task E	2	H	R
Sub-task F	2	H	H
$\sum_i st_{i\|r=3}$		6	6
$\sum_i st_i$		13	8
AVI		46%	75%

software interfaces, protocols and programs. The tools developed for this purpose are described in Sect. 4.3.3.

The general KPI of the system to provide sustainable, effective support for human operators is to display an AVI $> 75\%$ for most of the production tasks (e.g., every time a new pallet is presented to the operator). The soundness of the index relies on three main points:

- the swap of the task between human and robot is possible, and the sub-task performance is preserved, as stated in the above technical sessions.
- the planing allows in real-time the best allocation to reduce the index according to the real conditions (i.e., able to react to changing situations).
- the safety, the easiness, and the usability are always granted (as shown in the Section above).

4.4 Conclusion

The human-robot collaboration in Load/Unload stations has been considered. Team PIROS approached this challenge by adopting a modular solution. Actions have been modelled as function blocks. A scheduler assigns these actions to the human or the robot in order to optimize the Added Value Index (AVI), that is to assign sub-tasks with high added value to the human as much as possible.

The optimization of the AVI required high adaptability of the cell. This level of adaptability has been obtained by: (*i*) improving the manipulation abilities of robot, allowing the swapping of low added values activities between human and robot; (*ii*) realizing a human-aware motion planner which takes into account the human presence to plan the trajectory and to estimate the possible time exection variability; (*iii*) interacting with the operator by means of a gesture recognition; (*iv*) adopting a function block modular architecture to replan the task which requires a minimum effort of the operator; (*v*) developing a model-based Computer-Aided Risk

Assessment to support the designer of safety solutions during the phases of hazard identification, risk estimation, and risk evaluation.

Experimental results show the effectiveness of the proposed approach in a true industrial scenario.

The PIROS end-user is very satisfied of the human-robot collaboration framework designed during the project. In fact, he is interested to extend this paradigm to other cells of the plant, as well as to redesign the shopfloor flows to adapt them to reorganized layout.

For this reason, some members of Team PIROS are involved in the European Project Sharework [47] and in the *Assuring Autonomy* Project Recoll [48]. Within these projects, the considered LUS cell will reorganized by dividing the assembly and inspection of the tombstone from the mounting of the pallet of the FMS. This will improve ergonomics and the shared workspace between the robot and the human.

The human-robot collaboration in LUS handling and assembly is a challenging scenario, which implies the integration of many software and hardware components as well as designers skills. As Team PIROS, the main lessons we learnt is the required paradigm shift to convert a LUS cell to a human-robot collaborative cell. In fact, the collaborative robotic framework is not an add-on component which can be mounted on a non-automated cell without requiring changes. In fact, the same task imposes different timing and spatial constraints when performed by a human or a robot. Moreover, in many cells there could be a workspace limitation that does not allow the robot and the human to perform parallel tasks in proximity.

The implementation of a human-robot collaboration should always be used to rethink the way a cell works. In fact, a layout reorganization (w.r.t a simple addition of the robot in the preexisted scenario) improves ergonomics, the sub-task performance, and the number of tasks that the robot and human can make in parallel. Moreover, it can allow the designer to improve the material workflow.

References

1. Krüger, J., Wang, L., Verl, A., Bauernhansl, T., Carpanzano, E., Makris, S., Fleischer, J., Reinhart, G., Franke, J., Pellegrinelli, S.: Innovative control of assembly systems and lines. CIRP Ann. **66**(2), 707–730 (2017)
2. Chryssolouris, G.: Manufacturing Systems: Theory and Practice. Springer Science & Business Media (2013)
3. Tolio, T.: Design of Flexible Production Systems. Springer (2008)
4. Brettel, M., Klein, M., Friederichsen, N.: The relevance of manufacturing flexibility in the context of industrie 4.0. Proc. CIRP **41**, 105–110 (2016). Research and Innovation in Manufacturing: Key Enabling Technologies for the Factories of the Future - Proceedings of the 48th CIRP Conference on Manufacturing Systems
5. ElMaraghy, Hoda A.: Flexible and reconfigurable manufacturing systems paradigms. Int. J. Flex. Manuf. Syst. **17**(4), 261–276 (2005). Oct
6. Roveda, L., Vicentini, F., Tosatti, L.M.: Deformation-tracking impedance control in interaction with uncertain environments. In: 2013 IEEE/RSJ International Confererence on Intelligent Robots and Systems (IROS), pp. 1992–1997. IEEE (2013)

7. Siciliano, Bruno, Villani, Luigi: Robot force control, 1st edn. Kluwer Academic Publishers, Norwell, MA, USA (2000)
8. Geerinck, T., Colon, E., Berrabah, S.A., Cauwerts, K., Sahli, H.: Tele-robot with shared autonomy: distributed navigation development framework. Integr. Comput.-Aided Eng. **13**(4), 329–345 (2006)
9. Hu, X., Zeigler, B.P.: A simulation-based virtual environment to study cooperative robotic systems. Integ. Comput.-Aided Eng. **12**(4), 353–367 (2005)
10. Tsarouchi, Panagiota, Makris, Sotiris, Chryssolouris, George: On a human and dual-arm robot task planning method. Proc. CIRP **57**, 551–555 (2016)
11. Srivastava, S., Fang, E., Riano, L., Chitnis, R., Russell, S., Abbeel, P.: Combined task and motion planning through an extensible planner-independent interface layer. In: 2014 IEEE international conference on robotics and automation (ICRA), pp. 639–646. IEEE (2014)
12. De Silva, L., Lallement, R., Alami, R.: The hatp hierarchical planner: formalisation and an initial study of its usability and practicality. In: 2015 IEEE/RSJ international conference on intelligent robots and systems (IROS), pp. 6465–6472. IEEE (2015)
13. Pellegrinelli, S., Moro, F.L., Pedrocchi, N., Tosatti, L.M., Tolio, T.: A probabilistic approach to workspace sharing for human—robot cooperation in assembly tasks. CIRP Ann. **65**(1), 57–60 (2016)
14. Pellegrinelli, Stefania, Pedrocchi, Nicola: Estimation of robot execution time for close proximity human-robot collaboration. Integ. Comput.-Aided Eng. **25**(1), 81–96 (2017). Dec
15. Ding, H., Schipper, M., Matthias, B.: Collaborative behavior design of industrial robots for multiple human-robot collaboration. In: IEEE ISR 2013, pp. 1–6. IEEE (2013)
16. Ding, H., Heyn, J., Matthias, B., Staab, H.: Structured collaborative behavior of industrial robots in mixed human-robot environments. In: 2013 IEEE international conference on automation science and engineering (CASE), pp. 1101–1106. IEEE (2013)
17. Weitschat, R., Ehrensperger, J., Maier, M., Aschemann, H.: Safe and efficient human-robot collaboration part i: estimation of human arm motions. In: 2018 IEEE international conference on robotics and automation (ICRA), pp. 1993–1999. IEEE (2018)
18. Nicolis, D., Zanchettin, A.M., Rocco, P.: Human intention estimation based on neural networks for enhanced collaboration with robots. In: 2018 IEEE/RSJ international conference on intelligent robots and systems (IROS), pp. 1326–1333. IEEE (2018)
19. Thomas, U., Hirzinger, G., Rumpe, B., Schulze, C., Wortmann, A.: A new skill based robot programming language using uml/p statecharts. In: 2013 IEEE international conference on robotics and automation (ICRA), pp. 461–466. IEEE (2013)
20. Steck, A., Schlegel, C.: Managing execution variants in task coordination by exploiting design-time models at run-time. In: 2011 IEEE/RSJ international conference on intelligent robots and systems (IROS), pp. 2064–2069. IEEE (2011)
21. Brugali, D., Scandurra, P.: Component-based robotic engineering (part i)[tutorial]. IEEE Robotics Autom. Mag. **16**(4), 84–96 (2009)
22. Brugali, Davide, Shakhimardanov, Azamat: Component-based robotic engineering (part ii). IEEE Robotics Autom. Mag. **17**(1), 100–112 (2010)
23. Lewis, R.: Modelling Control Systems Using IEC 61499: Applying Function Blocks to Distributed Systems. Number 59. IET (2001)
24. Vyatkin, Valeriy: The IEC 61499 standard and its semantics. IEEE Ind. Electron. Mag. **3**(4), 40–48 (2009)
25. Strasser, T., Rooker, M., Ebenhofer, G., Zoitl, A., Sunder, C., Valentini, A., Martel, A.: Structuring of large scale distributed control programs with IEC 61499 subapplications and a hierarchical plant structure model. In: 2008 IEEE international conference on emerging technologies and factory automation, pp. 934–941. IEEE (2008)
26. Sunder, C., Zoitl, A., Strasser, T., Favre-Bulle, B.: Intuitive control engineering for mechatronic components in distributed automation systems based on the reference model of IEC 61499. In: INDIN'05. 2005 3rd IEEE international conference on industrial informatics, 2005, pp. 50–55. IEEE (2005)

27. Wenger, M., Eisenmenger, W., Neugschwandtner, G., Schneider, B., Zoitl, A.: A model based engineering tool for ros component compositioning, configuration and generation of deployment information. In: 2016 IEEE 21st international conference on emerging technologies and factory automation (ETFA), pp. 1–8. IEEE (2016)
28. Zander, S., Heppner, G., Neugschwandtner, G., Awad, R., Essinger, M., Ahmed, N.: A model-driven engineering approach for ROS using ontological semantics. arXiv preprint arXiv:1601.03998 (2016)
29. Allen, J.F.: Maintaining knowledge about temporal intervals. In: Readings in qualitative reasoning about physical systems, pp. 361–372. Elsevier (1990)
30. Iannacci, N., Giussani, M., Vicentini, F., Tosatti, L.M.: Robotic cell work-flow management through an IEC 61499-ROS architecture. In: 2016 IEEE 21st international conference on emerging technologies and factory automation (ETFA), pp. 1–7. IEEE (2016)
31. Marvel, J.A., Falco, J., Marstio, I.: Characterizing task-based human—robot collaboration safety in manufacturing. IEEE Trans. Syst. Man Cybern Syst **45**(2), 260–275 (2014)
32. Mayer, M.C.,Orlandini, A., Umbrico, A.: Planning and execution with flexible timelines: a formal account. Acta Inform. **53**(6–8), 649–680 (2016)
33. Alami, R., Clodic, A., Montreuil, V., Sisbot, E.A., Chatila, R.: Task planning for human-robot interaction. In: Proceedings of the 2005 joint conference on smart objects and ambient intelligence: innovative context-aware services: usages and technologies, pp. 81–85. ACM (2005)
34. Cesta, A., Orlandini, A., Bernardi, G., Umbrico, A.: Towards a planning-based framework for symbiotic human-robot collaboration. In: 2016 IEEE 21st international conference on emerging technologies and factory automation (ETFA), pp. 1–8. IEEE (2016)
35. Furia, C.A., Mandrioli, D., Morzenti, A., Rossi, M.: Modeling Time in Computing. Monographs in Theoretical Computer Science. An EATCS Series. Springer (2012)
36. PoliMi.: Zot: a bounded satisfiability checker. http://github.com/fm-polimi/zot (2012)
37. Pradella, M., Morzenti, M., San Pietro, P.: Bounded satisfiability checking of metric temporal logic specifications. ACM TOSEM **22**(3), 20, 1–20, 54 (2013)
38. Baresi, L., Pourhashem Kallehbasti, M.M., Rossi, M.: Efficient scalable verification of LTL specifications. In: Proceedings of the 37th international conference on software engineering, pp. 711–721. IEEE Press (2015)
39. Furia, C.A., Mandrioli, D., Morzenti, A., Rossi, M.: Modeling time in computing: a taxonomy and a comparative survey. ACM Comput. Surv. **42**(2), 6, 1–6, 59 (2010)
40. Askarpour, M., Mandrioli, D., Rossi, M., Vicentini, F.: Modeling operator behavior in the safety analysis of collaborative robotic applications. In: SAFECOMP, pp. 89–104 (2017)
41. Askarpour, Mehrnoosh, Mandrioli, Dino, Rossi, Matteo, Vicentini, Federico: Formal model of human erroneous behavior for safety analysis in collaborative robotics. Robotics Comput.-Integ. Manuf. **57**, 465–476 (2019)
42. ISO/TS 15066.: Robots and Robotic Devices—Collaborative Robots. International Organization for Standardization, Geneva, Switzerland (2016)
43. ISO 12100.: Safety of Machinery—General Principles for Design—Risk Assessment and Risk Reduction. International Organization for Standardization, Geneva, Switzerland (2010)
44. ISO/TR 14121-2.: Safety of Machinery—Risk Assessment—Part 2: Practical Guidance and Examples of Methods. International Organization for Standardization, Geneva, Switzerland (2012)
45. Askarpour, M.: Safer-HRC: a methodology for safety assessment through formal verification in human-robot collaboration. Doctoral dissertation, Politecnico di Milano (2018)
46. Lestingi, Livia, Longoni, Samuele: HRC-TEAM: A Model-driven Approach to Formal Verification and Deployment of Collaborative Robotic Applications. Msc, Politecnico di Milano (2017)

47. H2020 EC.: Sharework: safe and effective human-robot cooperation towards a better competiveness on current automation lack manufacturing processes. Founded from the european union's horizon 2020 research and innovation programme under grant agreement no. 820807. https://sharework-project.eu (2018)
48. "Assuring Autonomy". Recoll: Safety of reconfigurable collaborative robots for flexible manufacturing systems. Founded from "assuring autonomy" international program (by lloyds foundation and university of york). https://www.york.ac.uk/assuring-autonomy/projects/recoll/ (2018)

Chapter 5
FLA^{2}IR—FLexible Automotive Assembly with Industrial Co-workers

Georg Heppner, Felix Mauch, Stefan Scherzinger, David Timmermann, Pascal Becker, Stefan Ulbrich, Arne Rönnau, Peter Heiligensetzer and Fabian Fürst

Abstract This chapter gives an in-depth overview of the activities and technical approaches of the team FLA^{2}IR (FZI, MRK, OPEL) during the European Robotic Challenges to develop and deploy a flexible co-worker robot for the mounting of flexible polymer strips to a car door. The team won the EuRoC challenge with their extremely flexible approach that implements four key achievements. Fast and intuitive setup for new car variants: By developing the intuitive web based teach-in software CAD2Path, teaching times of under 5 min could be achieved. Flexible dexterous manipulation: By using a standard gripper with custom jaws and handling strategies a flexible polymer strip could be mounted successfully. Force based clip assembly: By adding an external force control loop, the industrial robot could push

G. Heppner (✉) · F. Mauch · S. Scherzinger · D. Timmermann · P. Becker · S. Ulbrich · A. Rönnau
Department of Interactive Diagnosis and Service Systems (IDS), FZI Research Center for Information Technology, Haid-und-Neu-Straße 10-14, 76131 Karlsruhe, Germany
e-mail: heppner@fzi.de

F. Mauch
e-mail: mauch@fzi.de

S. Scherzinger
e-mail: scherzin@fzi.de

D. Timmermann
e-mail: Timmermann@fzi.de

P. Becker
e-mail: Pascal.Becker@fzi.de

A. Rönnau
e-mail: roennau@fzi.de

P. Heiligensetzer
MRK-Systeme GmbH, Stätzlinger Straße 70, 86165 Augsburg, Germany
e-mail: Peter.Heiligensetzer@mrk-systeme.de

F. Fürst
Department of Advanced Manufacturing Technologies - Paint Shop and General Assembly, Opel Automobile GmbH, Bahnhofsplatz 1, IPC:42-05, 65423 Rüsselsheim am, Main, Germany
e-mail: fabian.fuerst@opel-vauxhall.com

F. Caccavale et al. (eds.), *Bringing Innovative Robotic Technologies from Research Labs to Industrial End-users*, Springer Tracts in Advanced Robotics 136,
https://doi.org/10.1007/978-3-030-34507-5_5

in clips "blind" by feeling the operation. HRC safety concept: The safety was validated by plant safety, even though ROS was used as control authority. The chapter describes the different teaching methods that were developed in the early stages and how the team finally solved the challenge and implemented the various aspects to tackle the extremely hard to automate problem of flexible polymer handling.

Keywords Industrial robot · Robotics · HRC · Flexible manufacturing · ROS · Force control · Flexible polymer handling · EuRoC · CAD2Path · Intuitive programming · Automotive

5.1 Introduction

While robots have become common in many productions and especially in the automotive sector, they still often follow old paradigms when it comes to programming. This leaves them unable to cope with increasing demands for a flexible production. Flexibility in this case refers to three different areas:

- Flexible setup and modifications of the robot programming
- Flexible execution of the task in coexistence with human workers
- Flexible handling of materials to cope with difficult tasks

Rapid changes in production like the introduction of a new product require robots to be easily programmable and adapt to new usecases. This becomes increasingly important when working with a synchronized production line where the reprogramming of a robot can impact the whole assembly line or if the robot is otherwise the single available resource for a specific production. Another important aspect of a more flexible production is the ability to perform cooperative applications where humans and robots work together or at least in close proximity. By removing the fences around robots, they can be placed more freely increasing the floor usage and travel ways of the goods. By intelligently partitioning the work between robot and humans tasks can furthermore be made more efficient when compared to pure manual or completely automated solutions. Finally, while traditional programming can solve assembly tasks with rigid parts such as chassis components, adaptive handling is required when flexible or very specialized materials are involved. By enabling robots to handle complicated parts and cope with more complex tasks, they can be used in a wider variety of tasks and the assembly of more products, which is opening up whole new manufacturing possibilities and in turn increases profitability of robotics.

FLA^{2}IR—Flexible Automotive Assembly with Industrial Co-Workers (Fig. 5.1) addresses all of these issues, bringing flexibility onto the shop floor of automated car manufacturing. Rather than replacing human work, this flexible assistant acts as a co-worker that performs repetitive, strenuous or dangerous tasks, leaving the humans free to perform the ones that would require an enormous effort to automate but are easily performed by a human. By focusing on novel methods FLA^{2}IR is able

Fig. 5.1 FLA^{2}IR setup at the Opel plant where it mounts car door sealings. The doors are transported on a hanging conveyor and stopped in front of the robot by a mechanical stop. Two different door types are used to show the quick adaptability of the system. The ROS controlled KUKA KR16 is equipped with an electric gripper with custom jaws to handle the flexible polymer strips (hanging over the doors). The fences around the application are a requirement of the plant for any prototype application but are not considered for the safety system of the application

to automate processes that were previously not possible to automate at all, pushing the boundaries of what robotics can do in a modern manufacturing environment.

During EuRoC development, we focused on key enablers for novel technology and how we can bring them into the application today and therefore create a significant impact in certain aspects of the flexible assembly. By developing an intuitive web based teach in of robot tasks we were able to reduce teach in times for a complex tasks to minutes and show that no prior knowledge is required to teach complex paths. A hybrid safety approach combines the benefits of currently available technology to make cooperation flexible while relying on proven technology to verify the workers safety. Finally a novel controller concept allows the use of force and impedance control on any robot. Combined with the mechatronic solutions for the gripper, we could show that the handling of flexible polymer strips is feasible, an area previously not possible to automate without specialized equipment.

The Chapter is organized as follows: Sect. 5.2 introduces the use case that was the ultimate goal for the developments of FLA^{2}IR. Section 5.3 details the stages of development and the technical approaches to the individual challenges we had to solve. Finally, Sect. 5.4 reflects on the achieved results and its impact to the automated assembly in general as well as giving some direction as to what development to pursue in the future.

5.2 Use Case Description and Motivation

For the field test stage of the project the task of **attaching a flexible polymer sealing to a car door** was selected (see Fig. 5.2). This task is currently done completely manually in the production line of the team member Opel and was considered to be unsuitable for automation up to the EuRoC challenge.

The door sealing or weather strip is clipped to the car door body where it prohibits water entry and reduces the cabin noises. The strip is a flexible part (EPDM polymer) and fixed with a varying amount of plastic clips depending on the door type. Zafira and Insignia door variants used in our case consisted of 35 to 39 clips which come pre-inserted into the polymer strip from another factory. While the pin positions are fix, their orientation is only roughly perpendicular to the sealing and can vary for several degrees without any pressure and even turn completely sideways (parallel to the sealing) when pressure is applied. Sealings are closed loops that fit the outer length of the door perfectly leaving no space once all pins are assembled. Each pin is spaced at a custom distance to the previous one and at some points the operator has to actively stretch the sealing to reach the next pinhole.

The process was selected mainly due to the fact that no automation solution could be found for this or similar tasks. It is well defined in its nature, yet it requires a set of skills that are often used in production and currently lacking in automation solutions.

5.2.1 Manual Process

The operator takes the strip from a supply rack and starts with the first clip which is defined and marked with color on the strip. The first mounting hole is also defined and known to the operator due to the clip layout of the sealing. One clip after another must be fixed manually (clipping directions always perpendicular to the strip surface) by pushing the pin deep enough into the pinhole. As the pinhole is slightly smaller than the pin diameter, two plastic hooks on the pin first get pushed in and then extend

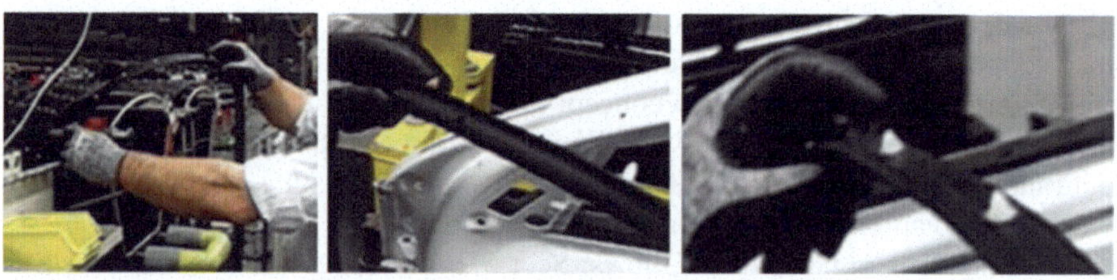

Fig. 5.2 Polymer sealing mounting at Opel plant. Sealings are grasped from a rack (left) with the pins facing upwards. The worker finds the first pin by color and aligns that with the car (middle). Subsequent pins are localized by the worker moving his hand along the sealing and feeling them. They are then attached in the following holes (right)

when inserted deep enough, locking the pin in place. Even though the mounting process itself is always the same much variation due to different door styles being assembled on the same production line can occur. For example front and rear doors with different door shapes and different number of mounting clips as well as different clip layout have to be handled by the operator on the same production line. In addition, different car variants (again front and rear door) are a possible. A quality check of the mounted strip including the correct clip position and a check if all clips are pushed in completely is required and currently done by the operator. The operator is using a visual check of strip to check for any obvious damage and a tactile check or sound feedback to ensure properly clipped clips. When correctly inserted the pins create a distinct clicking noise and the sealing can not be lifted of again when moderate pulling force is applied. The assembly process usually takes place at the moving line. Due to this it is fairly time critical process with an approximate time frame of 60 s for a complete assembly including the quality check at the end. As the car door is very large, this requires the operator to work in many different poses, which results in ergonomically difficult positions for the operator.

5.2.2 Automated Process

With current state of the art technology the automation of the assembly process is very challenging. Systems capable of handling flexible parts are in most cases limited to specific applications and often use very specialized tools. In combination with the different variants the skill of handling flexible components needs not only to cope with the material but adapt to the different scenarios as well. While the usage of sensors to abstract single human sensory skills is fairly common the required amount and combination of skills in our use case e.g. for finding the correct start clip of the sealing, the "feel" for the right amount of force to press in a clip or the quality check at the end of the mounting process is not trivial and, especially in the production environment, can not be solved with off-the-shelf available components.

Our proposed solution is to partially automate the assembly process to transform the manual assembly process into a HRC process with advanced flexibility. This means to automate the parts of the process which are time consuming, error prone and would require the operator to work in ergonomically bad positions while enabling safe co-existence with human operators which can still handle the tasks that require superior cognitive abilities. Four key challenges have to be addressed by the automated process:

- **Fast setup for new variants**:
  All poses of the holes have to be known by the system in order to insert each clip at the right position. These positions vary with each door variant. Teaching new door types is a tedious task that usually requires several days by an expert. The goal was therefore to develop a system that could be used by non experts to quickly integrate now product variants and adapt existing ones.

- **Flexible dexterous manipulation**:
  The sealings have to be handled in a way that the correct clip is at the correct orientation when attempting to insert a clip. For this, the non-uniform clip positions along the flexible sealing have to be detected and grasped. Additionally, the robot must stay in close proximity to the door as regrasping the seal is very hard and moving to far away from the door with the sealing in gripper will remove already inserted pins. The goal was therefore to develop a gripper solution capable of detecting pin positions, sliding along the sealing and grasping it firmly when required in conjunction with contour based trajectories close to the door surface.
- **Force based clip insertion**:
  The pins have to be inserted into the holes with enough force to secure them permanently, compensating all tolerances and the flexible pin position itself. As the pins would bend to the side when not aligned correctly, such a task requires a 'feeling' of the process. Additionally, it is desired to automate the quality control of the insertion, which is done by pulling the sealing up and testing if it comes loose. The goal was therefore to extend the industrial robot with a robust force sensing and control, making it truly adaptive.
- **HRC safety concept**:
  Current available safety certified approaches for industrial robots are not very suitable to be used in a HRC application. The HRC specific robots however are not suitable for the task due their size, speed and force restrictions. We therefore wanted to create a hybrid solution that can rely on flexible state of the art methods while still achieving safety certification at the plant.

The human operators handle the task of resupplying the robot with sealings, inserting it in such a way that the starting pin is aligned with the gripper by which a time consuming detection, alignment and regrasping can be avoided. Additionally, they perform the final quality control and correct any pins that were detected to be improperly inserted. One of our main goals was to develop a flexible solution which could at least be partially reused for a different application. This meant to use as much off-the-shelf products as possible, e.g. use specific handling strategies instead of a specialized gripper.

5.3 Activities and Results

EuRoC, as mentioned in the first chapters of this book, was organized in multiple stages, namely Simulation, Benchmarking, Freestyle, Showcase and Pilot experiments. The Simulation stage and Benchmarking stage presented the same challenge to all teams alike. In the following stages, the teams could set their own goals, however it was not required that they build onto another. The use case as described in the previous chapter was addressed in the final two stages, Showcase and Pilot experiments. Team FLA^{2}IR tried to spread the developments of capabilities for the final use case over all the stages and reuse as many developments as possible (see

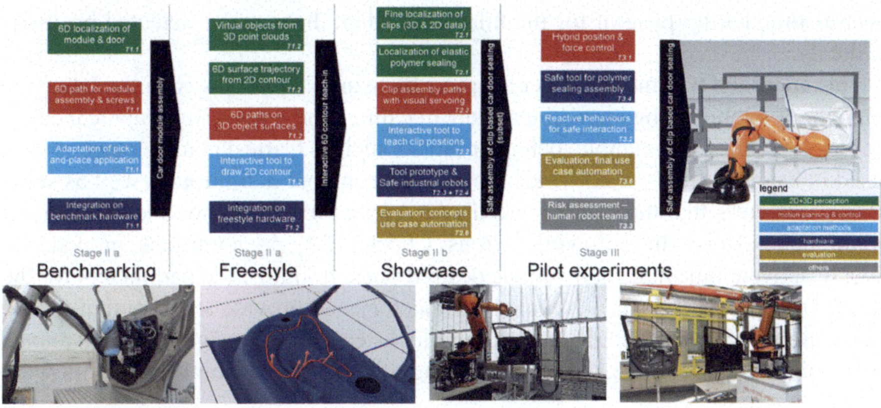

Fig. 5.3 Developments of team FLA^{2}IR in the different EuRoC stages. In Benchmarking a given assembly scenario had to be handled which resulted in quick adaption methods. In the Freestyle phase custom goals could be defined which resulted in various methods for intuitive teaching. In the Showcase phase the general application that was selected as pilot experiment was realized at laboratory conditions. Implementations in this phase focused on handling the polymers and clips. Finally the pilot experiment needed to prove the suitability of the implementation under plant conditions putting reactive behaviors and human robot collaboration as well as efficiency of the assembly at the top priority

Fig. 5.3). In recent years, the Robot Operating System (ROS) [1] has developed from a community-centered movement to a mature framework and quasi standard, providing a rich set of powerful tools for robot engineers, working in a large number of different domains. We therefore decided from the start, that ROS would be used as middleware for all developments which enabled rapid reuse and reconfiguration of developed solutions.

5.3.1 Development in Different Stages

In **Benchmarking** an assembly scenario was given where a car door inlay had to be inserted into a door and fixed in place with screws. A main challenge was that any part location (door, inlay, screws) could be changed (within a limit) and had to be detected as first step. All further assembly steps therefore had to adapt to the new part positions, making purely taught solutions impossible. This led not only to the development of perception methods for the parts, but more importantly improved the adaptability of the FZI Motion Pipeline significantly. By consequently using ROS TFs[1] for any operation and defining all movements and operations based thereon it could be used to create complex, yet fixed, trajectories, for example the approach

[1]TF is the ROS system to keep track of coordinate systems and their relation to another over time.

towards the pick up pose of for the inlay, and adapt them to the detected positions on the fly.

This approach was made further adaptive by using FlexBe, a system to define the behavior of a robot using hierarchical state machines and a powerful graphical editor for development. For example, behavior blocks such as Cartesian movement or joint based movement where used which would expect a goal definition as well as some other parameters like the desired speed or reference frame to use. Such a behavior would then perform the following actions: Check safety preconditions, unload previously active controllers, activate the desired ones, initialize the parameters for the the interpolator (speed etc.), execute the trajectory and provide status output for the action. Being able to reuse these building blocks easily was a significant advantage when building up a complex use case and a significant starting point for the following stages.

Freestyle was the first stage that allowed the teams to set their own (peer reviewed) goals. Team FLA^{2}IR chose to use this stage to focus on teaching methods for complex 6D surface trajectories and their force based execution with a special focus on rapid adaptation to new workpieces. The fast setup of trajectories tackles the first key challenge of the automated sealing assembly.

For the teach-in, 3 different methods where evaluated:

- **Force based contour learning**
  The user would, at the minimum, define a start and endpoint of a desired trajectory along the objects surface (the point could be taught 'in air'). In a learning phase the robot would then activate a custom force controller that applied a constant force (10 N) perpendicular to the surface of the object and move towards the taught goal point in a straight line. By maintaining a constant perpendicular surface force the orientation and position of the end effector adjusted automatically during this movement. Recorded poses are then used to update a spline interpolation of the path, effectively learning the surface contour of the object. In a reproduction phase the learned trajectory can be played back at higher speeds.
- **Force based path adaptation**
  As extension of the force based learning method a direct adaptation method was tested as second approach focusing more on user interaction. The user could simply grab and move the end effector during a learning phase. An impedance controller would try to keep the end effector on its previously learned path, making it easy to adjust small details but increasingly difficult to move it further from the taught path. While moving, new support points where recorded and used to update the learned spline for the trajectory.
- **CAD-2-Path**
  The objects CAD data was used in a web based software where the user could 'draw' support points onto the 3D model. By projecting them and subsequent sample points in between onto the objects surface a spline is created that accurately follows the surface. This trajectory could then be exported to be used in the motion pipeline backend. As CAD-2-Path is the tool that was further used it will be described in more detail in the following sections.

All methods were designed to cope with the sudden change of the target object. This was tested by teaching a trajectory on one car door and then changing it for a second door variant. The exploration method would compensate for changes by repeating the learning step or simply by adhering to the force constraints during execution. The CAD-2-Path methods would project the previously drawn trajectory onto the surface of the new object, recalculating the final robot path without altering the initial trajectory (while giving the opportunity to do so).

In the **Showcase** stage, the Opel use case was addressed with a focus on the second and third challenge of the automated process. Relying heavily on rapid prototyping, a custom gripper jaw design was developed that fulfilled the requirements of handling the sealing both by gripping it tight as well as letting it slide freely. Additionally a sensor concept was integrated into the gripper to detect the pin positions along the sealing. Clip positions at the door were taught by adapting the CAD-2-Path tool to handle holes.

While the initial idea was to use a visual approach for the clip insertion the team opted to extend the development of the force and impedance controller which enabled the robot to 'feel' the insertion. Another change to the initial plan was the safety system of the robot. While initially a full protective skin for close collaboration was planned, a more traditional approach using laserscanners was implemented following a shipping accident in which the complete skin was lost. This approach however, was ultimately closer to the requirements of Opel as not a close cooperation but rather a co-existence was requested.

During the **Pilot experiment** stage the focus was set on the fourth challenge of making the application save to use and able to work on the shop floor. The hybrid safety approach was further expanded during this phase and extended by a concept for the gripper safety and collaboration with a human operator, removing the need to initially grab the sealing. Another focus was to improve the speed and overall assembly process enabling end users to work with the FLA^{2}IR system.

5.3.2 Intuitive Teach-In Tool for Creating Complex Robotic Tasks

Teaching new trajectories is still a major hurdle regarding time consumption as well its complexity. Different solutions have come up in the past decades to overcome these problems, but only a few are commonly used in the industrial robotics environment. Nearly all existing methods require a trained programmer to teach new trajectories using the vendor specific programming languages, teach pendants or custom solutions. For different robotic manufactures different programming environments and languages have to be used resulting in a further specialization of the programming experts as well as a vendor lock in of the end user. One of the most time consuming parts of teaching a new trajectory is keeping the correct orientation and precise distances when following a surface, a common requirement for tasks

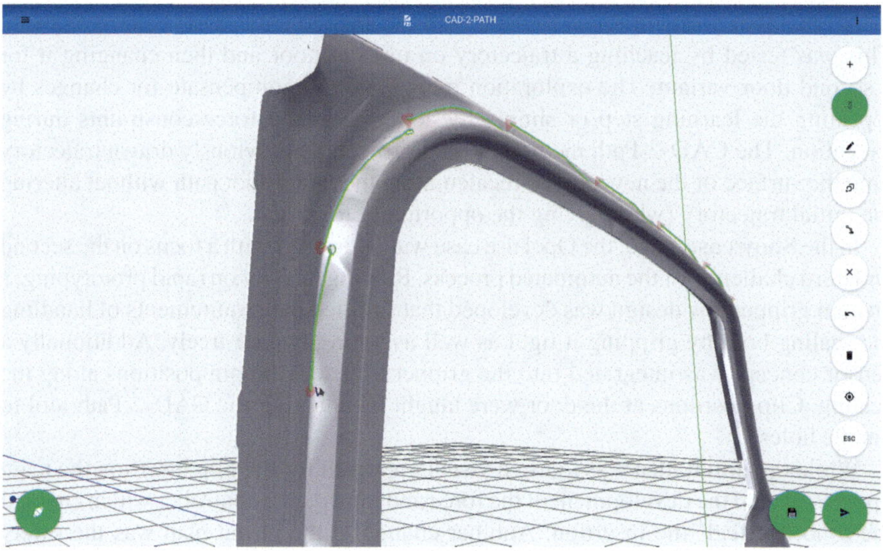

Fig. 5.4 The CAD-2-Path tool is a web based solution for fast and intuitive teach in. The user can 'draw' support points onto a 3D object created from the CAD data of the the target object. A spline is fitted between the support points but kept on the surface of the object. Various tools (such as move, insert support point, undo, etc.) as well as a project management and configuration handling make the tool very versatile and easy to use for different use cases

such as welding or the proposed handling of the polymer sealing. The first step to solve the use case was therefore to develop an easy and intuitive to use method for teaching a robot allowing even untrained people to use the tool effectively. To this end, the CAD-2-Path tool (shown in Fig. 5.4) was developed.

The developed solution is web based to allow usage on a wide range of devices from desktop computers to smartphones or tablets. We used the material-design and the angular framework as a basis for the graphical user interface and combined it with libraries such as D3js for the 3D viewer and the roslibjs to interact with the ROS based backend. The core functionality such as the projection onto the object, building the cubic spline representation and the tools to modify the trajectory were implemented in typescript which is compiled to javascript. By using a pure web based approach to the interface, we impose very little restrictions on the system running it and allow the end user to stay in a familiar environment.

The generated trajectories are not specific for our framework but are Cartesian trajectories usable for different robots and setups. They are given relative to the coordinate frame of the object (intersection of horizontal and vertical line in Fig. 5.5) and can be further defined by setting various options within the tool. These options include the name of the reference frame for export, an offset towards the surface of the object or the resolution of subsampling points when exporting the trajectory. CAD-2-Path provides an integrated calibration tool which can be used in junction

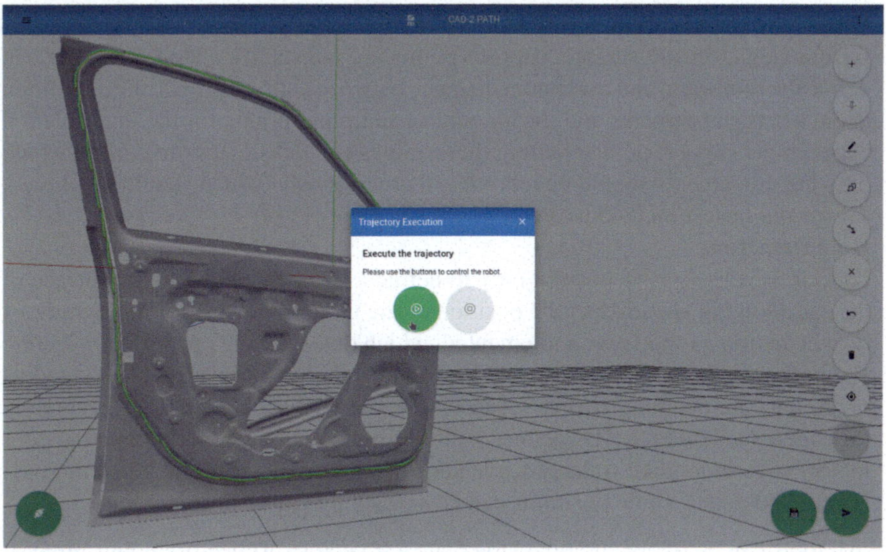

Fig. 5.5 Execution of a newly taught trajectory can be started or stopped directly form the CAD-2-Path tool making it the main user interface the user has to interact with

with a teach panel on the used robot. The user indicates 3 reference positions in the model and a reference name. A ROS based calibration tool is then used to record the tooltip position of the robot when the user indicates that it is at one of the 3 calibration points. After 3 positions have been stored the pose of the workpiece can be calculated and is published in relation to the robot base. During execution this reference is used to calculate the actual trajectory in relation to the robot base and in turn the joint values of the axes. As the trajectory is only defined by the support points and interpolated by a cubic spline, the export resolution can be set to an arbitrary number of points, allowing for a much denser sampling during execution than during creation. For EuRoC we combined the CAD-2-Path tool directly with our motion pipeline and process control to start the mounting application from the GUI (see Fig. 5.5).

To teach a new trajectory the tool requires a mesh file of the object or scene in which the trajectory shall be generated. Since for our use case we require a trajectory on the surface of the car door the tool matches the orientation and position of the trajectory points with the surface information gathered from the mesh model. A trajectory is created by the operator using different functionalities included in the tool:

- **Point-to-Point function**:
  The user selects different support points on the surface of the object. The resulting trajectory is a cubic spline based interpolation between these support points. Placing individual support points gives the user a precise control over fine aspects of the trajectory and to leverage the interpolation qualities of the cubic spline.

- **Free Drawing function**:
  By drawing over the surface, support points are constantly added. The user can define the minimum and maximum distance between support points, the maximum angle between segments and the amount of sampling points for the interpolation between the support points. Setting these values enables the user to control which drawing movements should be reproduced precisely and which should be filtered. This is the fastest method to create new trajectories.
- **Hole Detection**:
  For our use case we needed not only to define a trajectory close to the surface of the car door, but also extract the position of the mounting holes for the sealing. A hole detection method was implemented, which detects if the operator is selecting an opening in the model surface and tries to fit a circle within a preselected radius range into the opening. If this is successful, the center of the hole is accurately measured and added as a trajectory point. By indicating the holes around the door in sequence individual trajectories leading from one mounting point to the next are created and the hole positions can be extracted from their first point.

All tools can be freely combined to create a trajectory. Trajectory points are projected onto the model to match the curvature of the surface. Extra functionalities like moving, inserting or deleting a support point make changes to an existing trajectory easy. During the 'drawing' the trajectory is constantly evaluated and rendered providing immediate feedback which is important to get details such as the curvature right. If any problem is detected, e.g. the trajectory is traveling over a hole, the corresponding sub trajectory is marked as defective and the resulting trajectory is not accepted as a valid solution.

The results of the use case showed that through CAD-2-Path a user could teach a previously unknown door model in less than 15 min after only a short introduction to the system. Trained users such as FLA^{2}IR developers can perform a set-up in less than 5 min.

5.3.3 Manipulation of Flexible Polymer Sealings

Handling flexible objects is very hard for robots due to various reasons. The complexity of the describing physics behind deformations, limited force-sensitivity and dexterity of the manipulators and the complexity of integrating tactile and possibly visual feedback in the control algorithms are just some of the problems that have to be faced. While individual solutions to the problems have been shown to work, they are still at the edge of the state of the art and rarely combined in a convincing way.

Humans are particularly good in this domain which is one of the reasons for our end user Opel to heavily rely on human labor for many assembly tasks. Mounting flexible rubber sealings to car doors comprises most of the mentioned challenges. What adds to the difficulty in an industry environment is the demand for high robustness, which often rules out state of the art solutions from robotic research labs.

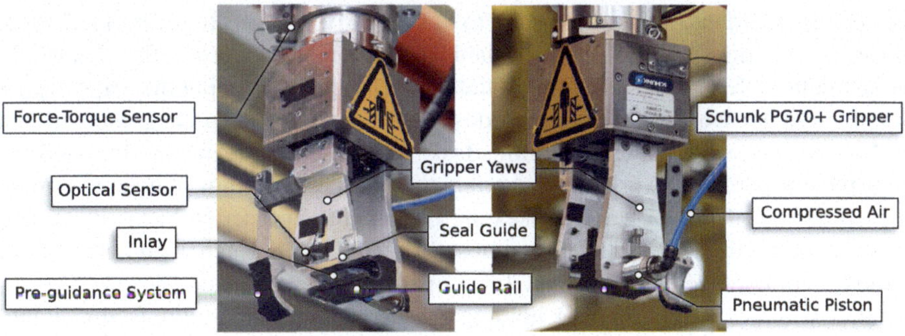

Fig. 5.6 The tool used to handle and mount the flexible polymer sealings. An industrial 2 finger gripper from Schunk was used and equipped with custom made jaws containing a 3D-Printed sealing guidance, an optical sensor for pin detection and a pneumatic piston for the actual pin insertion. By using standard hardware instead of highly specialized components we show the versatility of robotic manufacturing for many areas rather than solving a single problem with specialized and expensive mechanics

With a focus on real industrial requirements, such as reliability and cost efficiency as well as flexibility in terms of standard components in mind, we therefore developed our concept of handling the flexible polymer sealings by strongly relying on industrial proven components where possible. Instead of designing a complex, single use, specialized tool for the polymer handling we opted for using a standard gripper and focused on intelligent handling strategies and customized gripper jaws that can be easily manufactured. Using a standard gripper for handling the sealings limited us in dexterity, but also offered the opportunity to be close to industrial deployment. Figure 5.6 shows the standard gripper *Schunk PG70+* with its custom components, developed during the challenges.

When mounting the sealing, three actions have to be achieved by a tool: First, the sealing needs to be grasped tightly above or around the pin to guide it to its corresponding mounting point and push it in. In some cases, this might even require to stretch the sealing a little bit, which requires a very tight grip. Second, after a pin is inserted, the sealing has to be released to move to the second pin. Third, the next pin has to be identified by sensing as the distance between the pins is different for every pin and it might even be missing. While the first task is easily done, we quickly realized that letting the sealing go to reach the next pin is not an option. Not only would this require 3D sensing, but also a much more complex handling as the sealing can easily twist in every direction, resulting in complex regrasping strategies that require several planning steps. Furthermore, as the sealing is a fixed loop, there is simply no space left to safely grasp the sealing once most of the pins are set.

The challenge in designing the gripper was therefore to imitate two main aspects of the dexterous strategies that we humans apply. First, conveying the sealing safely inside our gripper—much like we humans do with our softly closed hands, running along the sealing and caging it with our fingers and palm, and second, approximating the effectiveness of an index finger for the actual push in of the clips. For the first part, we heavily relied on rapid prototyping to obtain early versions to support our tests. In various iterations, we designed mechanical components with chamfers that would guide the flexible sealings throughout the mounting process, using the gripper jaws to tighten or loosen the sealing when desired. Figure 5.6 shows the *guide rail* and *seal guide* that are considered with this task. As the sealing is mounted around the door, the gripper has to work in every orientation of the wrist joint. When the gripper is used on top of the the door the sealing that is not yet inserted hangs down and therefore meets the seal guides at an almost 90 degree angle which would often result in the loss or damage of the sealing. To compensate this, a pre-guidance system was installed which is acting as a hook holding the sealing up and therefore feeding it into the guidance rail at a much better angle. In order to remove the pre-guidance system once the complete sealing is mounted and to avoid any damage when pressing the sealing onto the door it is mounted using a flexible spring steel element and padded with protective foam.

For the second part, we used the whole robotic arm for pushing the clips in during the showcase stage, which was both slow and error prone. In the final Pilot experiment stage we therefore included a pneumatically driven piston that could directly apply pressure on the pin, imitating the usage of an index finger. This mechanism is shown in Fig. 5.7 with the support hose for compressed air and the pin as part of the flexible sealing. The air pressure, which we control through an electronic valve, pushes the inlay (See Fig. 5.6) mounted onto the piston outwards and in turn elevates the pin for about half a centimeter, which is sufficient for the clip to fix its position with small

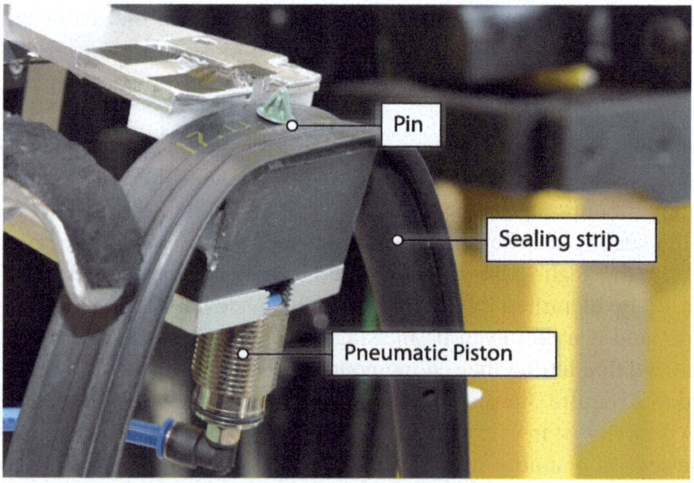

Fig. 5.7 The gripper with an inserted sealing strip. The sealing is fixed to the door with an initial pin and then moved through the gripper from left to right (in the shown orientation). When in the shown orientation the sealing glides easily into to guide rail, when turned 180 degrees, the pre-guidance system (left) feeds the sealing at an shallow angle. The pin is pushed into its mounting hole by extending the pneumatic piston and detected by the thin optical fiber located in the middle of the gripper

barbs in the car door's holes. By including a pneumatic piston mechanism into our manipulation system, we achieved a major speed-up for the process of pin-clipping in comparison to pure robot motion based solutions.

To detect the next pin passing by when conveying the sealing through the gripper we made use of an optical fiber sensor to detect the intensity of light reflections. Upon detecting a pin, we switch from a gliding mode to a firm pinch-like grasp for the air-driven assembly. The actual clipping is triggered upon detecting the right spot on the car doors with force-based search, as is described in the next section.

On the assembly lines at Opel, workers take the sealings for each door from a special supply rack. An automatic pick up from a sealing rack and re-adjusting was shown in the Showcase stage of the project, using a fixed environment tool for regrasping. Albeit being technically feasible, the automatic pick-up of the sealings required too much time and slowed down the application to that point, that deployment in a real assembly line was out of the question. We therefore opted to use this step to highlight the co-worker possibilities of the process by letting a human worker insert the sealing strip into the gripper at the start of the assembly.

5.3.4 Force-Sensitive Control with Assembly Strategies

During the EuRoC challenges, we quickly realized that we would need to describe applications both with end-effector motion and associated force profiles at the same time. This became apparent in the Benchmarking stage, where inserting the car door inlay required compliant motions to cope with the inlay hitting the door due to imperfect localization. *Impedance Control* is one of the solutions that comes close to meeting these requirements. However, albeit being a major solution for some lightweight robots, we could not expect industrial manipulators to provide such control interfaces in general. Moreover, in the industrial target environment of EuRoC, *the classic* industrial robots with high gain position servos, such as the KUKA KR16, still represent the majority of manipulators used today. This class of robots is often very restricted in its interfaces as the manufacturers control them tightly. Existing interfaces are usually for setting feed forward joint trajectories $\boldsymbol{q}(t)$ with individual reference setpoints. The limited frequency with which users may implement an outer closed loop control and the fact that compliance is not inherently available, but must instead be imposed by controlling motion, make it very difficult to achieve force sensitive end-effector behavior and nearly impossible to move fast and stable in contacts with the environment.

What we could assume however, is that the robots can be equipped with sensors to measure the wrench $\boldsymbol{f}^s$ at their end-effector tools, and that they provide functionality to bias the sensors. Relying on additional sensors is not restrictive, given that 6-axis force-torque sensors have become widely available for industrial robots. In combination with our control concepts, they constitute an effective and cost efficient add-on solution to a wide range of robots.

To derive our control concept, we start with the common equations of motion for a rigid body system in joint coordinates $\boldsymbol{q}$

$$\boldsymbol{H}(\boldsymbol{q})\ddot{q} + \boldsymbol{C}(\boldsymbol{q}, \dot{\boldsymbol{q}}) = \boldsymbol{\tau}_u + \boldsymbol{J}^T \boldsymbol{f}^{ext} \tag{5.1}$$

where $\boldsymbol{H}$ denotes the symmetric, positive joint inertia matrix, $\boldsymbol{C}$ encompasses centrifugal and Coriolis terms, as well as gravitational forces, and $\boldsymbol{\tau}_u$ denote the motor torques in the joints of the manipulator. External forces and torques $\boldsymbol{f}^{ext}$, acting on the end effector map to joint space with the joint Jacobian $\boldsymbol{J}$. Under the assumption that we move slowly in contacts and that the robots we consider already compensate for gravity in their lower level control, we omit $\boldsymbol{C}$ and $\boldsymbol{\tau}$, such that we obtain

$$\ddot{q} = \boldsymbol{H}^{-1}\boldsymbol{J}^T \boldsymbol{f}^{ext} \tag{5.2}$$

as an instruction to move according to external disturbances. While we let $\boldsymbol{J}$ reflect our real system, we generalize $\boldsymbol{H}$, by setting the masses of all links to some unit values. Note that it is not required to replicate the real system dynamics, since we deploy (5.2) as a *forward dynamics solver* to solve the inverse kinematics problem [2]. The system dynamics are already compensated for in the low-level joint position servo.

Figures 5.8 and 5.9 depict the control scheme for the Cartesian force and motion control respectively, in which $\boldsymbol{x}^d$ and $\boldsymbol{f}^d$ are the desired feed forward motion and desired wrench at each time step and $\boldsymbol{q}$ is the computed joint position setpoint for the robot.

Figure 5.10 is essentially the combination of the two. The three schemes highlight the *solver* of the inverse kinematics problem, using a *PD* regulator on the general control error $\boldsymbol{e}$, which regulates $\boldsymbol{f}^c$ according to $f^c = k_p e + k_d \dot{e}$ for each of the six Cartesian components. In all control schemes, the highlighted part in grey runs in a separate control cycle using the virtual model. The gains k_p and k_d are, in combination with the chosen dynamics of $\boldsymbol{H}$, a partly redundant means to adjust the systems responsiveness. In the case of pure force control, $\boldsymbol{f}^d$ is equilibrated by the

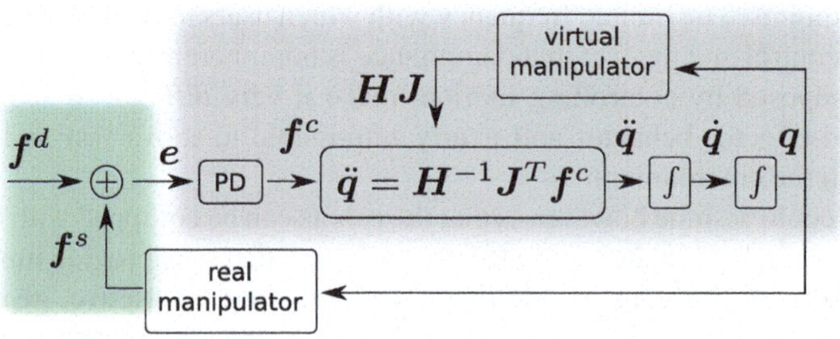

Fig. 5.8 Cartesian force control. The part highlighted in grey (on the right) runs in a separate control cycle using a virtual model

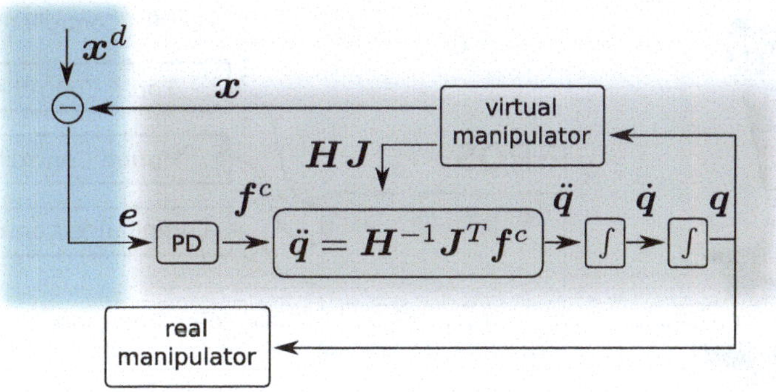

Fig. 5.9 Cartesian motion control. The part highlighted in grey (on the right) runs in a separate control cycle using a virtual model

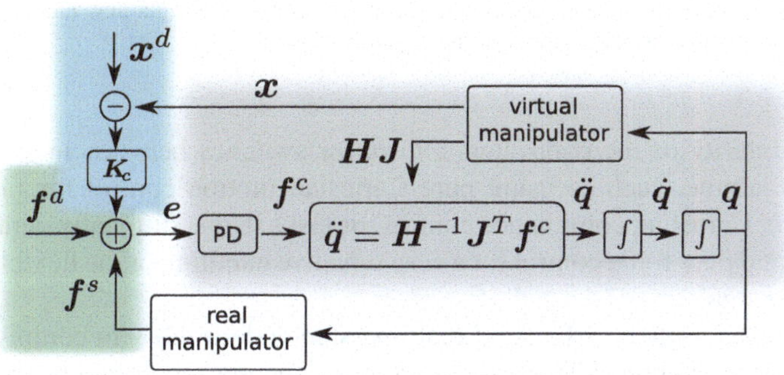

Fig. 5.10 Cartesian compliance control is essentially a combination of the two previous controllers. The part highlighted in grey (on the right) runs in a separate control cycle using a virtual model

individual measured components in f^s. For pure motion control, this is a feed forward regulator without any real system feedback, assuring fast and instantaneous motion. An advantage of using forward dynamics to solve the inverse kinematics problem in this context is the ability to interpolate between distant setpoints $\boldsymbol{x}(t)$. Unlike classical inverse kinematics solvers, such as search-based solutions, our approach does not only provide the final result for reaching the given target pose, but also provides the steps in between by lowering the gains k_p and k_d respectively. In the compliance controller, users can specify an additional Cartesian stiffness $\boldsymbol{K_c}$, which is independent of the controller gains.

As we use ROS for all EuRoC implementations we also decided to implement our Cartesian controllers from Figs. 5.8, 5.9 and 5.10 in the ROS control framework [3], meaning to build upon an established and tested architecture for handling and managing robot specific controllers in realtime. Additionally, it would make it easier for us to contribute to a large community in the later dissemination phase. A simplified version of the architecture is depicted in Fig. 5.11.

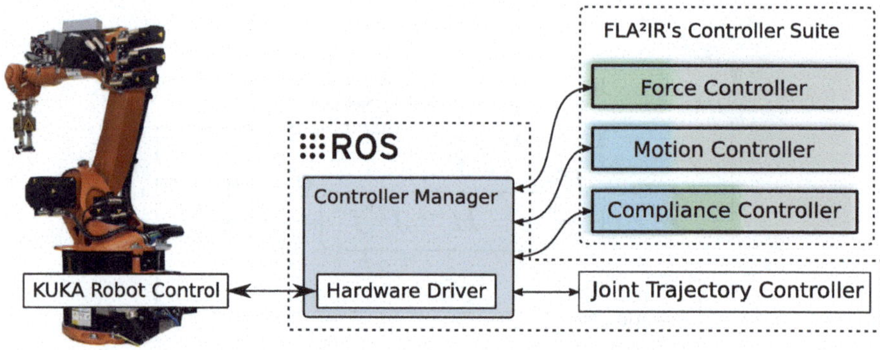

Fig. 5.11 Control architecture with individual components. The hardware receives joint commands from the hardware specific interface loaded into the controller manager. Different higher level controllers are loaded by the manager and activated based on the current process requirements. The structure makes it easy to change either the hardware, by implementing a new hardware interface or adding a new controller independent of the actual hardware

The modularity of the controllers allows for switches between individual controllers in realtime, such as using pure Cartesian motion control for fast motion in free space, Cartesian compliance control for tasks that additionally demand force sensitivity, and pure force control for a very sensitive handling of the flexible sealings during pin search.

All code was written as fast and realtime-safe code in C++ to comply with the ROS control architecture. These realtime requirements are common in industrial scenarios, and will be important for a successful dissemination of the software. Both the controllers and strategies are designed to be reused in a wider range of applications, posing little requirements to the robots used for the tasks.

5.3.4.1 Assembly Strategy

We decided to not include vision sensors into the manipulation pipeline of the flexible door sealings for multiple reasons. Adding vision would require the end user to acquire additional, costly hardware, and maintain fixed lighting and visibility conditions in their plants to produce accurate results. Additionally, detecting features on the flexible black sealing or the small holes on colored reflective surface of the car door are non trivial problems that would require extensive development and training data to reach a reasonable reliability level. The more decisive reason, however, was that it proved to be very difficult to take serious advantage of vision sensors to enhance the in-contact manipulation strategies: Fitting the flexible sealing to the car door nearly perfectly occludes the process of pin clipping in contact. During manual assembly on the lines, human workers use their sight for a rough estimation of the pin clip positions. Their visual servoing, however, is likewise limited due to occlusions by the sealing itself, leaving the workers to rely on their tactile sense for haptic feedback.

When implementing the manipulation strategy in the pilot, we assumed a localization component on the real conveyor belts in manufacturing, which we imitated during test runs with a manual calibration (see CAD-2-Path). Once the door was localized, the clip positions are known as they were previously taught in the CAD-2-path tool. Due to mechanical play, the localization of the door itself is however only accurate up to roughly ±2 mm. Additionally, due to the construction of the pins, their orientation, while mostly perpendicular to the sealing, is not fixed, resulting in a few degrees of uncertainty which can not easily be detected by any kind of sensor. The orientation offset will lead to another 1–3 mm offset of the pin tip. The combined offsets are more than enough to miss the mounting hole with the tip of the pin when relying purely on position information. As a consequence, we chose to fuse the approximate hole positions with readings from the force-torque sensor as feedback, using the compliance controller from Fig. 5.10 and search strategies in contact. Figure 5.12 shows the usage of the controllers.

We used pure motion control for movements between the holes, relying on the approximate Cartesian paths as indicated with the CAD-2-Path tool by the users. The path is stopped after a fixed distance once the optical fiber detects the pin moving by, resulting in a position where the sealing pin is placed right in the middle of the gripper. At this point the gripper was closed to grasp the sealing tightly and then moved to the theoretic mounting hole position. As the motion was controlled with the pure motion controller the sealing could easily be stretched to reach to

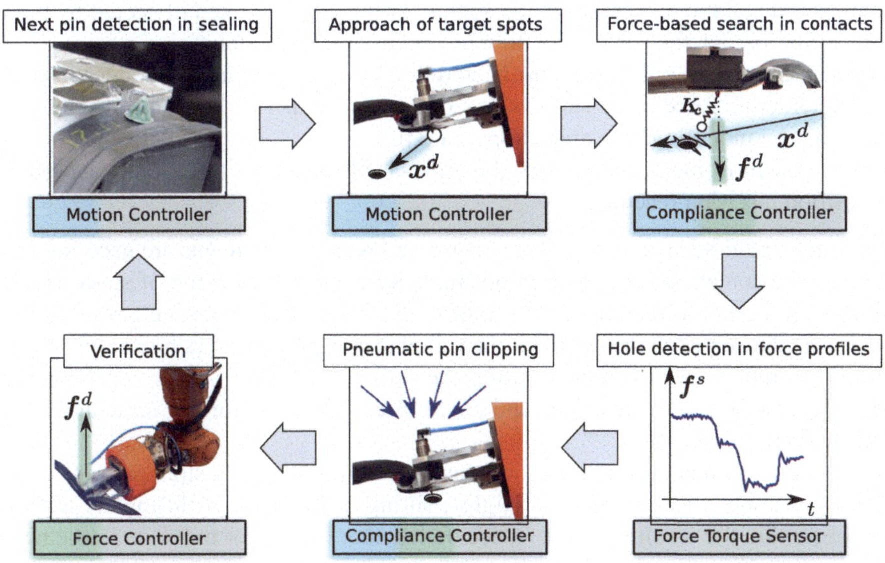

Fig. 5.12 Steps of pin clipping with cartesian controllers

insertion position. Forces that act onto the sealing at this point were removed as bias from the sensor readings which removed their influence for the following step. We then switched to compliance control as we had to anticipate collisions in the close proximity of estimated hole positions. Surface normal localization errors were compensated by applying a downwards oriented force $\boldsymbol{f}^d$, and assured by a sensitive *surface hovering* with little pressure. Orthogonal to this axis, we applied search strategies in form of $\boldsymbol{x}^d$, which we decoded as partly random motion with respect to the estimated hole poses and an exploration margin of few millimeters that were validated empirically. Through asymmetrically adjusting the stiffness $\boldsymbol{K}_c$ of the end-effector, we both followed this exploring motion with a certain stiffness and applied a sensitive pushing force to the pin of the sealing. It is imperative to keep the force $\boldsymbol{f}^d$ at a perfect balance during this procedure. When set to low, the pin will hover smoothly over the surface, but also have not enough force to move into the mounting hole once it is found as the stiffer controller for the search axis pull it out immediately. When set to high, the pin rests secured in the mounting hole if it is hit on the first try but can be bent so far to the side when missing it that a successful insertion becomes impossible.

Once in searching mode, we continuously observed the force profile from the readings of our force torque sensor, and deployed a hand-crafted detector to determine when to trigger the pneumatic piston. The successful insertion is decided on multiple criteria that mirror the procedure performed by a human:

- Detection starts, once the target force of $\boldsymbol{f}^d$ has been reached, meaning the pin is in contact with the surface.
- When the theoretical position is met, the insertion depth has reached a certain threshold and the movement in the search directions is blocked (detected by high reaction forces), the perfect position has been hit immediately.
- Otherwise the search movement is started and observed until a sudden drop of counteracting force in the surface direction is detected. This happens once the pin has hit the mounting hole and 'drops' into it.
- When the movement in the search directions is blocked after this event, the perfect position has been reached.

The gate conditions to watch for the constrained search movements are necessary as the pins are sometimes mounted in positions where less than 1 mm of space is left to move in certain directions before hitting the door. After many iterations of the detection, the approach proved to be very reliable except for total failure cases where the pin would be bend to much during the initial touch down. This was however only the case with defective pins and could in the future be further mitigated with detection strategies.

Similar to human performance, we checked our pin insertion strategies and their desired outcome, i.e. a mounted and tightly sitting sealing strip, by lifting the sealing and observing the resulting force to decide whether to retry or move on to the next clip.

5.3.5 Safety Concept

The safety of a robotic application is handled with a classical approach in most factories. Industrial robots are surrounded by fences and any violation of the defined safety parameters (e.g. opening the gate) result in an emergency stop of the system providing a clear separation between robot and human operator. For EuRoC we aimed to present an alternative approach to allow flexible collaboration and especially co-existence of human and robot.

To this end, we demonstrated the captive and tactile robotic skin shown in Fig. 5.13 developed by the project partner MRK. The sensor data is evaluated in a safety certified SPS and passed on to the control computer running ROS. The capacitive data is interpreted by the ROS system. If any approaching human is detected it will try to ramp down the robot to a complete stop but not abort the overall control flow. The execution of the process is paused by the currently active state itself as all states where extended to react to a pause signal. As soon as the human leaves the sensor field, the system resumes the executed trajectory and the idle process by removing the pause signal. If the tactile sensor of the skin is triggered the robot safety SPS signals the controller to execute an emergency stop.

Fig. 5.13 The used Kuka KR16 equppied with the robotic skin developed by MRK Systeme. A capacitive system detects proximity and halts the robot. If the skin is deformed by touching it, a safe emergency stop is triggered

The idea of this safety concept was to enable close collaboration between the robot an human co-worker. But during discussions with Opel it became clear that such close collaboration would not be favored in this use case.

The finally implemented safety concept therefore is a hybrid approach with a more conservative focus that nonetheless enables safe co-existence of human and robot in a shared workspace, making the robot overall move flexible. Similar to the two zone approach with the skin laserscanners with a certified safety field and a non-safe warning field were used to supervise the area around the robotic cell. Inside the safety field a conventional CE certifiable safety concept is used that brings the robotic system into a hard emergency stop as soon as anything enters the safety field while the robot is moving. This conventional approach results in a stop of the whole mounting process. After manually re-enabling the robot, the user drives the robot to a safe pose and resets the system. While this method leads to a robotic cell that can be safety certified, such a system does not allow human-robot collaboration (HRC) like inserting the sealing into the gripper.

To prevent an emergency stop when a human approaches, a soft safety stop has been introduced using the ROS Industrial stack and the laserscanner's warn fields. As soon as the warning field is violated the robotic system is ramped down to a stopped state. If this can be achieved before the inner safety field is violated, the robot can be transferred into a deactivated state in which the robot's stand still is monitored by the certified safety system inside the robot controller. Additionally, the gripper power is disconnected and the pressure is removed from the pneumatic system. In this stage it is save to approach the robot, as any violation to the deactivated state would lead to an immediate emergency stop. An overview of the robotic cell layout with the corresponding safety zones is shown in Fig. 5.14.

In contrast to conventional safety methods the robot can recover from such a deactivated state as soon as the safety field is cleared. The robot can then continue the process as soon as the warning field is cleared as well. This enables human-robot collaboration while still providing a certified safety system. The developed safety mechanism represents a close coupling of conventional safety technology implemented by our industry system integrator MRK and an intelligent process management implemented by FZI. While being developed for this particular use case this concept can be used for any robotic application that requires hard safety concepts with occasional human-robot interaction. In particular, the ability to combine the unsafe ROS System, which offers huge flexibility benefits, with a safety approach that passes plant inspection is a huge potential for any future application.

For the gripper we developed a similar concept like the one of the robotic skin. Since the gripper is no HRC certified product but just a simple electronic parallel gripper, we showed how such a standard component can be extended with additional components to be suited for a collaborative task between human and robot. The gripper is surrounded by a plastic armor equipped with tactile and capacitive sensors. While the capacitive signal is used to ramp down the system it is also possible to perform complex movement adaptations to avoid any contact with the human worker. The tactile sensor would lead to an emergency stop including the cut of

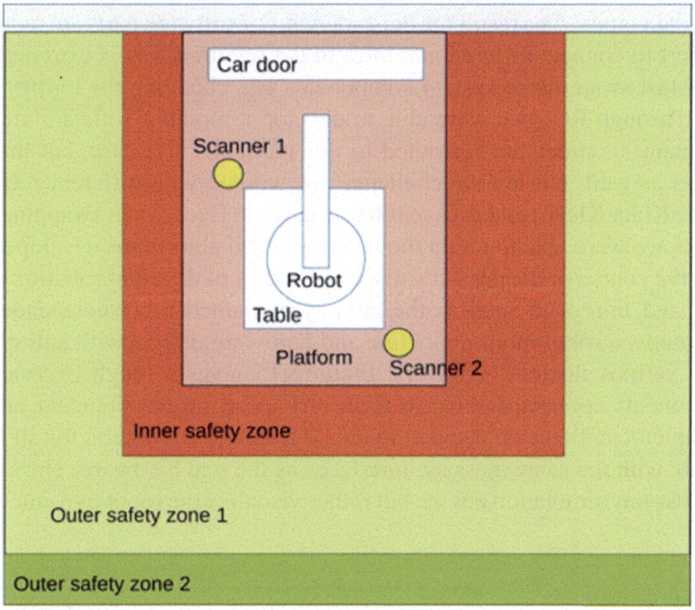

Fig. 5.14 Sketch of the robotic cell layout with marked safety zones. The laserscanners monitor the complete area around the robot. When any object enters the inner safety zone, a safe stopped detection in the robot controller is activated which would trigger an emergency stop in case of any robot movement. The outer safety zone 1 if monitored with the non safe warning field which triggers a ROS controlled ramp down of the robot and a halt of the overall process. Objects in the outer safety zone 2 do not influence the process

the gripper power and the opening of the secure valve to remove pressure from the pneumatic system. Ultimately, this additional safety approach was not strictly needed in combination with the laserscanner approach.

5.3.6 System Architecture—Flexible, Modular, Powerful

The process management developed for the Use Case stage is a good example of combining the power of conventional industry automation, the ROS Industrial software stack and a modern, user-friendly interface.

The software architecture we developed during the project is based around the robot operation system(ROS) [1]. ROS is an open source framework used by many robotic researchers. It offers core functionalities like inter module communication, component discovery and package abstraction and generally acts as middleware. ROS strongly encourages the approach of creating a modular software solution instead of a strictly monolithic one. This is supported by the publish/subscribe communication approach of ROS which uses a or remote procedure call style communication with abstracted data types. As a result the dependencies of software parts are reduced. The needed effort to connect different modules in the required way is extremely simple and allows fast swap out of system components e.g. changing the torque sensor or the robot. Through ROS we were able to develop a modular software architecture that uses many modules not restricted to our particular use case, but many other applications as well. During the challenge we worked with different robots (First UR10, later Kuka KR16) and different force sensors. Even when swapping out core components we were able to retain the core logic and algorithms developed so far.

During the course of the project software modules of different robotic skills were developed and improved such as the already explained robot-independent-force-control strategy, a robot motion pipeline and hardware drivers with halting support as well as various flexible behaviors. Figure 5.15 gives a rough overview of the core components connected through ROS. To speed up development and enable parallel implementations we decided to create a simple simulation for all hardware components with the same message interfaces as the real hardware. The simulation would not use any simulation engine but rather visualize the robot movement in tools

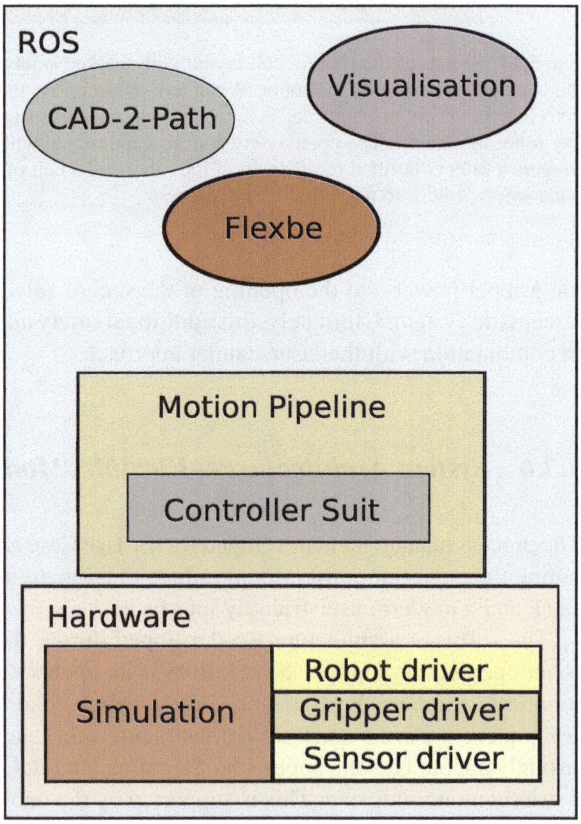

Fig. 5.15 A overview of the used core components

such as RVIZ. Custom interfaces enabled the injection of sensor values for example to test the force based detection strategies.

For the robot driver we used and extended the already existing Kuka RSI driver present in the ROS Industrial software stack. Our motion pipeline with the control suit mentioned in a previous sections was used to execute the Cartesian trajectories from CAD-2-Path as well as the joint based utility trajectories as moving to a home position. The overall application control is handled in FlexBE, a state machine package available for ROS. To inform the operator about the current and next actions of the system we visualized the robot and its target as well as some additional information by using rviz.

By using FlexBE we were able to create a highly flexible overall process control. FlexBE offers the option to define states and behaviors. States are simple functionalities, like activating a valve, while behaviors are a collection of states to provide a bigger functionality e.g. pushing in the clip force based. To prevent behaviors of becoming too big and complex it is possible to nest behaviors into each other. One of these behaviors is the SET_CLIP_COMPLIANT behavior, shown in Fig. 5.16 used to insert a clip of the sealing into a hole of the door with the previous described compliance controller. As it can be seen the behavior itself is rather simple with well defined state transitions. The behavior is triggered from higher level behavior, or for testing and development purposes can be triggered stand alone. Transition between different states or behaviors can be assigned with varying autonomy levels which

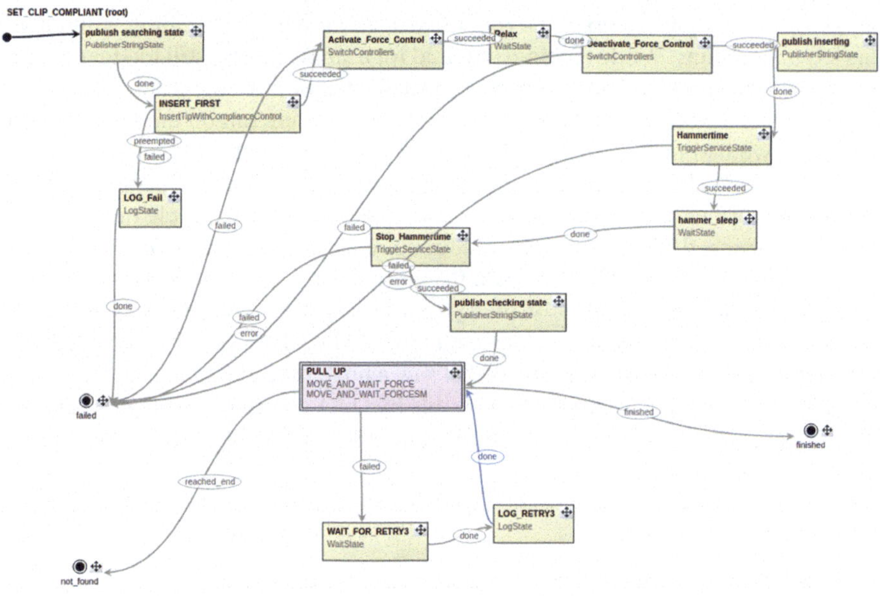

Fig. 5.16 The FlexBe behavior to insert a clip using the compliance controller. Various states are combined to form a more complex behavior. Clear state transitions make the behavior understandable while user data provides flexibility at runtime

enables easy semi automatic testing of behaviors as they would execute autonomously until reaching a transition specifically set to a lower level which requires user input. It is also possible to define specific outcomes of the behavior like in this example: finished, failed and not_found. User-data is not fixed in FlexBE as it can be set and modified by using ROS functionalities. The behavior concept offers a nice way to quickly swap out complete parts of the software by just replacing a behavior with a new one with the same inputs and outputs, e.g. for testing a new push in behavior.

The system's flexibility could be shown by adapting to a new door type in less than 15 min by a non-expert with a very good outcome. By using ROS and our developed flexible behaviors only changes to the user-data were required, leaving the underlying software structure untouched.

5.4 Conclusion

We have set out to build a flexible automotive co-worker solution and presented the pilot application of mounting flexible sealing strips to a car door, a task previously not lending to automation at all. The developed system has proven its robustness with inserting 28 out of 35 clips for the Insignia door in 11:01 minutes and 37 out of 39 clips for the Zafira model in 12:03 minutes during the official evaluation run. In these metrics it should be mentioned that the high amount of unset pins on the Insignia door were caused by the team having inserted the wrong clips into the sealing. With the correct brown pins inserted the robot managed to mount the sealing with only 3 pins left in later runs. While the time required for a single door is still too far off to be used in the synchronized line, this performance is a big improvement compared to the Showcase stage where it took about 45 min for a complete run on the Insignia door and where we had to skip 6 clips deliberately due to the risk of losing the sealing from the gripper. It could be expected that an additional time reduction could be achieved by further optimizing, however, not to the extend that a cycle time of about 60 s can be reached. Such a speed would require a pin to be set in about a second, which is unlikely to be achievable with the current force based strategy.

In user studies the usability of the system was verified and received an overall positive feedback from the users. Especially the CAD-2-Path tool was received with much praise from different specialists that saw a huge potential in it. The CE certification of the system has been proven possible by Opel's work safety department. All CE relevant requirements had to be fulfilled in order to get clearance for the system to run within the plant.

From a technical perspective the following core competencies were developed and verified in this use case:

- **A hybrid safety concept for safe human robot interaction**
  The used laserscanner zones provide two important safety functions: By using the non-safe fields to disable the software in a coordinated way, the efficiency of the application and the ease of use are greatly enhanced. On the other hand the safe

outputs, which activate the robot integrated safe-stop-check, provide the required level of safety to ensure CE-level verifiability. Thus the application showcases the concept for an efficient, yet usable safety concept for industrial robots using ROS.

- **Easy, intuitive, robot independent teach-in tool CAD-2-Path**
  The developed tool CAD-2-Path offers a novel and intuitive solution to a common bottleneck of robotics, the programming of complex tasks especially in close contact to a workpiece. The solution was already verified in a number of demonstrations and shown to work for untrained users in real environments without introduction. The application can easily be extended to fit various needs like different robot-tools and adapted to every workflow (online control or offline trajectories) with ease. The FZI will most likely use this method in several projects putting it close to product level with the potential to be used in many applications.
- **Add-On force control for industrial robots**
  The methods used for the pin insertion required a new approach to force-based control. As the developed method is robot independent and provides state of the art force control it is a result with a very high impact. It enables automation of previously inaccessible tasks such as the sealing assembly, opening whole new areas of applications. The developed method will most likely be applied in prototypes rather than production as it currently still requires slow robot motions in contact. However, the ability to use it offers many exciting options.
- **Flexible polymer handling**
  The developed concept of using specialized gripper jaws on an otherwise standard industrial setup has proven to work and enabled the robot to safely handle flexible polymer sealings in analogy to a human counterpart. While not reaching the ambitious cycle times, the realization of this previously non automatable approach highlights the flexibility of the used technologies.

During the EuRoC challenge, team FLA^{2}IR managed to score high results in every stage and ultimately was selected as winner of the overall challenge. The developed system provided a good mix of state of the art to tackle novel problems, proven technology to increase the robustness and engineering to tie the results together in a flexible and reusable way and was therefore able to show the possibilities of a truly flexible assembly co-worker.

Unfortunately, the implemented pilot application will, in its current form, not be used in production for several reasons.

First, the business case is not convincing enough for the automotive sector. The current solution relies on a human worker to cooperate with the robot. Implementing it would enhance the working conditions and free some of the workers time up to do other tasks, but could not replace him completely. Even with a full automation, only a single production step would be automated which provides a to little return of investment to justify the robot costs. To optimize the value gained, the robotic solution needs to be part of a broader restructuring.

Second, the restrictive cycle time of about 60 s in the synchronized production line prohibits the use of flexible adaption methods that simply require more time. Even when setting the pins without force control and perfect accuracy, the cycle time results in roughly 1–2 s per pin which is barely enough to set individual pins in the way currently proposed. To reach such times specialized tools or processes, like gluing instead if clipping, could be used. Another option is the change from a synchronized production cell to an asynchronous step that could simply take more time. As such a changes would greatly affect the overall process they are not taken lightly and have to be evaluated carefully in terms of cost, quality and process integration.

Third, the current layout and organization of the assembly make the usage of robots difficult in general. When using the robot without fences, the overall layout of the factory becomes more flexible, but has to allow for more free space in order to implement the required safety distances. Also, as the current layouts and material flows are optimized for human workers, making it difficult to plan robotic backup solutions to ensure a 100% reliability.

While these reasons prohibit the usage of the application at the moment, they might be mitigated in the future. When, for example, more than one application is modified to be automated or the assembly process is reworked in general, the proposed solution might become viable and provide a valuable building block for the complex assembly steps.

Of course, the application itself offers much potential to optimize. The current search strategies involve a random search. By investigating imitation learning with recurrent neural networks the time spent on searching could possibly be improved. Detecting the pin hole insertion could also be improved with a data driven approach in contrast to a hand-crafted detector. A more general force based detection might even be transferable to different application tasks. And finally, the slowest parts of the application could of course always be replaced with specialized sensors or mechanical solutions.

Although our solution as a whole is not yet straight forward to put into practice, individual components have proven to be very valuable. CAD-2-Path for example could save significant resources: In current Insignia models, Opel's high price cars, components including the main chassis, interior, doors and roofing contain in total more than 130 Meters of adhesive sealing, which is applied by robots. Programming the robots for this task consists of an off-line phase, where the initial robot motion is set-up, and an online phase, in which the application programmers correct hundreds of points in order to comply with limitations and fine tune the trajectories for tool collisions. Both phases last equally long. When using FLA^{2}IR's CAD-2-Path tool, Opel's application engineers were impressed by the speed-up with which they programmed these trajectories. While they didn't see CAD-2-Path saving them time in the online phase, they saw a massive improvement for the off-line preparations up to 50%. Apart from using CAD-2-Path in gluing and welding applications, it could also easily be modified to suit various other use cases. In the same way as the hole-search was implemented other shape detectors, tools and methods to interact with a complex trajectory could be implemented easily making it suitable for almost any tasks. The CAD models that are currently used could further be replaced with meshes generated from a laserscanner or similar sources, gearing the application more towards an online adaption, rather than offline preparation. Combination of live sensor acquisition and model data is another reasonable approach, making the tool useful for repair and maintenance tasks.

The proposed safety approach worked very well and can be used for other applications without too much effort opening up many possibilities. A significance advantage of the approach is the ability to use potentially unsafe systems such as ROS for applications with a heightened safety requirement today and in production environments. It would, of course, be beneficial if the used framework could provide the level of safety on its own. However, fulfilling the safety requirements usually prohibits many of the dynamic features a system requires to truly push the state of the art which is the reason why most systems today are either safe but very conservative or not safe at all. Combining the approaches in such a way offers a realistic way of adopting novel methods into real applications today.

The flexibility of the pilot demonstrator, in particular the motion pipeline for arbitrary motion execution and the FlexBe system for adaptive tasks are also of high value for both, specific applications as well as future developments. By having a robot independent option of generating robot movements we open up novel applications, foremost the assembly on a moving line as all of our movements can simply be set to a relative base. We have already shown this to be feasible in a pilot application after the use case using the same pipeline to mount a screw in a workpiece on a moving conveyor belt, a use case of high significance also for the team member Opel. Various other use cases implemented by FZI have used parts of the developed pipelines and could be significantly sped up due to the flexibility of the approach. The good experience with FlexBe is further extended by looking at behavior trees for even better reconfiguration of developed behaviors.

Finally, by adding force control to industrial manipulators we open up many applications previously impossible to automate and can seamlessly integrate such force control into the developed pipeline. Making a process compliant offers many advantages, especially when workpieces need to be placed accurately or inaccuracies have to be compensated for. While this is a mostly isolated problem in the optimized assembly lines of large factories today, the trend towards specialized items, smaller lot sizes and novel materials will require this feat more frequently in the near future. Having an add-on solution is a cost effective way of enabling existing factories to follow this trend.

A four-year project typically generates a rich amount of experience and expertise among the participants. EuRoC was not different, which is why we present an excerpt of our insights and lessons learned from this exciting undertaking.

Starting as a challenge, it was clear that EuRoC provided a different motor of motivation compared to *classic* European projects: Competition was a central element during the running project. We contestants had to continuously assert ourselves against other teams on different stages, not only in 'standardized' benchmarks, but also with respect to freely, self-chosen objectives, for which innovation and impact were as important factors as technological challenge. We felt that the competition lead to a positive and motivating atmosphere among the contestants, and advancing to the next stages was an interesting form of feedback of our work and achievements with the direct consequence of further funding. The scheme of funding for results, rather than writing a complex proposal and reports shifted the results from theory to practise. This lead to significant achievements which are much more exploitable than the usual scientific results of research projects on the one hand, but also solutions that employ much more engineering solutions than scientific ones on the other hand. While the competition format might therefore not be suited for scientific breakthroughs on fundamental levels, it proved to be a very valuable tool to bring current state of the art solutions to a usable TRL that directly benefits the targeted industries.

However, there were also more challenging aspects. Its sometimes difficult to squash our solutions into the given metrics, which had to make us contestants comparable in some way. But designing objectives on the other hand required us to examine our solutions in depth, which helped us in working strategically towards our own goals. While having a use case as driving motivation, reaching a productive state in a research project was overly ambitious. Although we had a strong use case within the German automotive industry, it remained more than challenging to achieve a positive business case with our end user Opel, regarding the required 60 s for a complete door sealing assembly. Further work will be necessary here for an integration into the assembly line, although we showed it to be in technical reach.

Being close to production level also required us to evaluate our systems with respect to ease of use and applicability, which are in contrast not always the focus in research oriented development. To this end, we conducted a study within a hands-on workshop to investigate important parts of our system (CAD-2-Path) with test users with different educational background, but all related to automotive assembly at Opel. The study participants prepared the software-sided setup of two door derivatives (Zafira and Insignia) by themselves and taught the entire path for assembling the door sealing on both doors. The participants then transferred the taught-in path to the robot which performed the task with a single tick on the start button. Apart from new hints and perspectives, it was especially rewarding for us to see novices to the field ‘program’ the robot for the sealing assembly task in minutes.

Finally, we address part of our success to building upon the Robot Operating System (ROS) as a powerful middleware and the strong and thriving community behind its development. Packages such as the *ROS-industrial* driver Kuka experimental, the *ROS control* framework and FlexBe are great examples for packages that enable quick focus on the actual challenges of an application, such as the polymer handling.

References

1. Quigley, M., Conley, K., Gerkey, B., Faust, J., Foote, T., Leibs, J., Wheeler, R., Ng, A.Y.: ROS: an open-source robot operating system. In: ICRA Workshop on Open Source Software (2009)
2. Scherzinger, S., Roennau, A., Dillmann, R.: Forward dynamics compliance control (FDCC): a new approach to cartesian compliance for robotic manipulators. In: 2017 IEEE/RSJ International Conference on Intelligent Robots and Systems (IROS), pp. 4568–4575 (2017)
3. Chitta, S., Marder-Eppstein, E., Meeussen, W., Pradeep, V., Tsouroukdissian, A.R., Bohren, J., Coleman, D., Magyar, B., Raiola, G., Ludtke, M., Perdomo, E.F.: ros_control: a generic and simple control framework for ROS. J. Open Source Softw. (2017)

Part III
Achievements in Challenge 2: Shop Floor Logistics and Manipulation

Introduction

Christian Ott

Collaborative robots have long been promoted by the academic robotics community as a technological approach to strengthen the European manufacturing industry in the challenges faced by novel flexible manufacturing paradigms such as Industry 4.0. The appearance of the KUKA Lightweight Robot IV as a commercial product, originally developed by the German Aerospace Center (DLR), can be seen as a turning point in the robotics industry. While traditionally robotics industry was dominated by stiff actuation concepts aiming for maximum positioning accuracy, a strong interest in soft robotics concepts and manipulators with controllable compliance arose. This development lead to a new generation of collaborative robots which allow for compliant physical interaction with the environment as well as safe interaction with humans. The vision is a flexible and intuitively programmable robotic companion which enables human–robot collaboration tasks in small and medium enterprises. In particular, safe physical human–robot interaction plays a central role in these systems. In addition to compliant motion control, this requires highly sensitive contact detection and reaction, human-aware motion planning, as well as robust perception of humans and the environment.

Nowadays, in fact, a large family of such collaborative robots are available on the market, e.g. KUKA iiwa, the Universal Robots UR 5 and UR10, as well as the recently introduced Franka Emika to name just a few. In most industrial use cases, these robots are currently used as collaborative manipulators within a dedicated work cell. The combination of mobility with dexterous manipulation can be seen as the next fundamental challenge for collaborative robots in industrial applications. While the research community has studied mobile manipulation from an academic point of view in great detail, the transfer of these concepts into successful industrial solutions is still at a very early stage. As identified in the Strategic Research Agenda of the Public–Private Partnership in Robotics (SPARC), modular, dynamic and flexible logistics and automation solution on the shop floor is one of the keys to the factory of the future. This motivated the focus of EuRoC's Challenge 2 on Shopfloor logistics and manipulation.

Differently from Challenge I, this Challenge was based on a prepared mobile manipulation system consisting of an omnidirectional platform equipped with a KUKA iiwa collaborative manipulator. The system was further extended by an onboard camera system, a gripper and a navigation unit. Together with the technology provider, KUKA, the Challenge Host, DLR, prepared the overall system and provided a convenient programming interface. All required basic functionalities for the operation of the system were established such that the Challengers could focus on their respective use cases without being burdened by low-level issues and maintenance of the system.

The participating teams considered use cases in a wide range of industrial domains, such as Aerospace, Automotive, Site/Facility Maintenance, as well as Logistics. The following chapters give a detailed presentation of the use cases, the main technical approach and the unique challenges the two Finalists in Challenge II faced.

Team RSAII, lead by IK4 Tekniker, presented a solution for a realistic warehouse logistics scenario proposed by the company ULMA. In this scenario, RSAII combined a stationary robot for order preparation with a mobile manipulator for order distribution. The use case involved a wide range of skills, including automatic recognition, grasping and manipulation of the different goods, motion and packaging planning, as well as reactions to human users in the environment.

The second finalist, team TIMAIRIS, presented a solution for feeding piles of blankets to an automatic packaging machine. This use case was proposed in connection with the end user IMA. The final demonstration involved the simultaneous feeding of two packaging machines with different types of piles. The team demonstrated online replanning based on the machines filling levels, skillful handling of the flexible piles, using a customized gripper, and gesture-based interaction with a human worker.

Both use cases required a tight integration between system mobility and dexterous manipulation skills, and thus addressed the next level of human–robot collaboration tasks.

Chapter 6
RSAII: Flexible Robotized Unitary Picking in Collaborative Environments for Order Preparation in Distribution Centers

Loreto Susperregi, Ane Fernandez, Jorge Molina, Ander Iriondo, Basilio Sierra, Elena Lazkano, Jose Maria Martinez Otzeta, Mikel Altuna, Leire Zubia and Unai Bautista

Abstract Distribution centers are facilities designed to store and manage different products to be redistributed to another location or directly to customers. Although these centers have a high degree of automation, tasks such as picking are hardly automatable and are one of the most labor-intensive and monotonous tasks in material handling operations. Picking, i.e. the process of collecting items to create a package for shipment, is the main source of errors and lack of efficiency. Although the fully

---

L. Susperregi · A. Fernandez · J. Molina · A. Iriondo (✉)
IK4-TEKNIKER, Iñaki Goenaga 5, Gipuzkoa, Spain
e-mail: ander.iriondo@tekniker.es

L. Susperregi
e-mail: loreto.susperregi@tekniker.es

A. Fernandez
e-mail: ane.fernandez@tekniker.es

J. Molina
e-mail: jorge.molina@tekniker.es

B. Sierra · E. Lazkano · J. M. Martinez Otzeta
EHU-UPV, Manuel Lardizabal, 1, 20018 Donostia, Spain
e-mail: b.sierra@ehu.eus

E. Lazkano
e-mail: e.lazkano@ehu.eus

J. M. Martinez Otzeta
e-mail: josemaria.martinezo@ehu.eus

M. Altuna · L. Zubia · U. Bautista
Ulma Handling Systems, Barrio Garagaltza 50, 20560 Oñati, Spain
e-mail: mialtuna@ulmahandling.com

L. Zubia
e-mail: lzubia@ulmahandling.com

U. Bautista
e-mail: ubautista@ulmahandling.com

F. Caccavale et al. (eds.), *Bringing Innovative Robotic Technologies from Research Labs to Industrial End-users*, Springer Tracts in Advanced Robotics 136,
https://doi.org/10.1007/978-3-030-34507-5_6

automation of the picking task is highly desirable, multiple factors such as the high variability of parts or industrial requirements (safety, robustness...) limit the fully automation of the task. RSAII developed a hybrid order picking approach, in which robots and humans share the same workspace, combining high automation, flexibility and safety. The objectives defined in the project addressed key technical and industrial challenges and three experiments were performed with increasing complexity: (1) Free-style: Mono-reference picking with high availability, (2) Showcase: Safe multi-reference integrated with current pick to light solutions, (3) Field test: Collaborative solution for Unitary Picking in order preparation area of a Distribution Center. As a result of the Field test, it has been developed a working prototype that works in a realistic environment and represents the foundation for real implementation of the system at ULMA customers' premises.

Keywords Mobile manipulation · Intralogistics · Bin-picking.

6.1 Introduction

A **Distribution Center** is a specialized facility designed to store and manage different products to be redistributed to another location or directly to customers. The main functions are receipt, temporary storage, and redistribute goods according to the customer orders as they are received. Activities associated to goods flow include transport, storage and handling for a wide variety of references. Distributions centers are an integral part of the order fulfillment process, especially for online retailers and e-commerce businesses (Fig. 6.1).

Today's goods flow faces several challenges: customers demand nearly 100% order accuracy; orders are smaller and more frequent; and productivity and low costs are in this sector a competitive advantage.

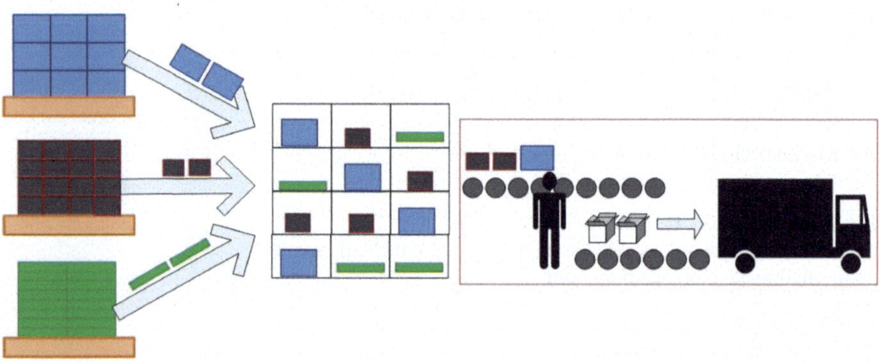

Fig. 6.1 Product flow in a distribution center

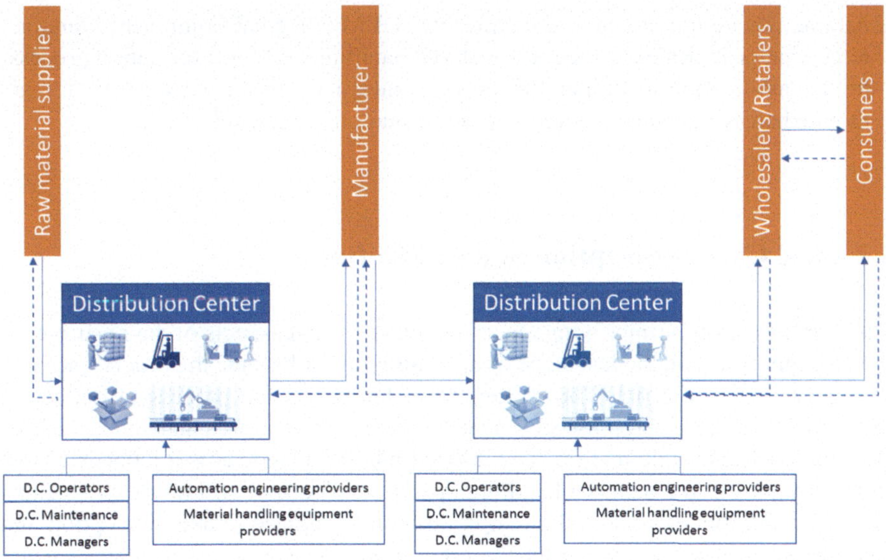

Fig. 6.2 Main stakeholders

Several stakeholders can affect or can be affected by the activities, objectives and strategies in the operation of a Distribution Center (D.C.). The needs and requirements of all them must be considered to provide valuable robotic and automation solutions. Figure 6.2 provides a description of the main stakeholders and the relationship among them:

- Customers for a D.C.: Goods stocked are redistributed to manufacturers, wholesalers, retailers, or directly to consumers. Material distribution policies may affect to product order preparation or return management.
- D.C. operators: There are different activities performed day to day by in a distribution center and several profiles can be considered: Operators perform direct labour in the D.C. (truck unloading and break down in pallets, inventory and tag of products, transport received pallets from the receiving dock to the storage racks, put products into racks, pull products from the racks, etc.); Managers perform support activities including information systems, considered as indirect labour and they include, among others, maintenance of buildings, inventory management (tracking and placement of products), inspection and acceptance of incoming and outbound product, etc.; Maintenance performs equipment routine maintenance and repair tasks.
- Automation solution provider is an engineering company that designs, builds, and supports logistics solutions (automated storage and picking solutions) that optimize material and information flow in a D.C. The solution provider with the OEM is also responsible to provide on-site support for installation, start-up and training of routine service and repair tasks of the system.

- Automation equipment provider represents OEM (Original Equipment Manufacturer) that provides equipment for material handling and for information processing for automation including conveyors, cranes, industrial trucks, AGV, storage, mobile barcode scanners, wireless printers and other handheld devices.

6.2 Use Case Description and Motivation

The Use Case aims at improving one of the most critical aspects of the Distribution Centers supply chain, the last stage of its automation solutions: the orders' picking, as represented in Fig. 6.3. The picking, also known as the process of collecting items to create a package for shipment, is the last and most challenging task of Distribution Center automation. The manual picking of goods is a labor-intensive and costly jobs in the material handling operation. The staff still holds 90% of this strenuous and monotonous task, as it is difficult to replace human perception and hand dexterity that can cope with the high variation of parts sizes, materials and

Fig. 6.3 Current intralogistics picking solution for a distribution center: Person to good (left) and Good to person configurations (right)

packaging methods. However manual picking and packing is the main source of errors and lack of efficiency.

Building automated solutions is highly desirable. However, the viability of a fully automated system is limited by aspects such as the high variability of items to be manipulated, the industrial requirements in terms of robustness, safety, real-time operation and the space constraints. Up to now, in Distribution Centers these limitations do not allow automated solutions rise. However, a **hybrid collaborative solution** can impact the market providing higher efficiency and availability rates comparing to current manual solutions. In recent years, there are many factors supporting the viability of more flexible automated solutions such as: the emergence of robot manipulators able to work more collaboratively without fences and the progress in perception systems.

The main idea behind the Use Case is to provide a hybrid order picking approach, in which robots and humans share the same working space, combining high automation with the needed flexibility and safety. At present, although highly desirable, the feasibility of a fully automated system is limited by aspects such as the high variability of items to be manipulated, the industrial requirements in terms of robustness, safety, real-time operation and the space constraints. The objective is to set up a working prototype of a hybrid automation solution in the picking area of a Distribution Center, where robots autonomously pick objects reacting safely to people position around them.

Currently, the work flow at the packaging area is as follows, workers manually load the elements that compose an order into boxes, a slow, expensive process that is shown in Fig. 6.4.

1. The automated storage system approaches a set of boxes to the order preparation area. Goods to be picked are coming in boxes with different configurations.
2. People in the packaging area pick objects of different references from the upper boxes as in Fig. 6.4.
3. Then the objects are placed by the operators in the packaging boxes, lower boxes in Fig. 6.4.

Clear advantages are foreseen from the partial or total automation of the manual picking process:

- Error reduction and associated cost reduction: some internal data in ULMA show that in current manual picking solutions up to 5% of order references have errors or defects. Typical errors of manual picking are: Counting error 2%, Reading error 23%, Mistaken item 30%, Missing item 45%.
- Cost reduction: as an expected impact an hybrid approach can make feasible reduce a person per turn, since the robot can reduce operator burden.
- Space reduction: traditional robotic solutions with fences for protection has a direct effect in the layout to be defined to build a robotic solution with associated cost.

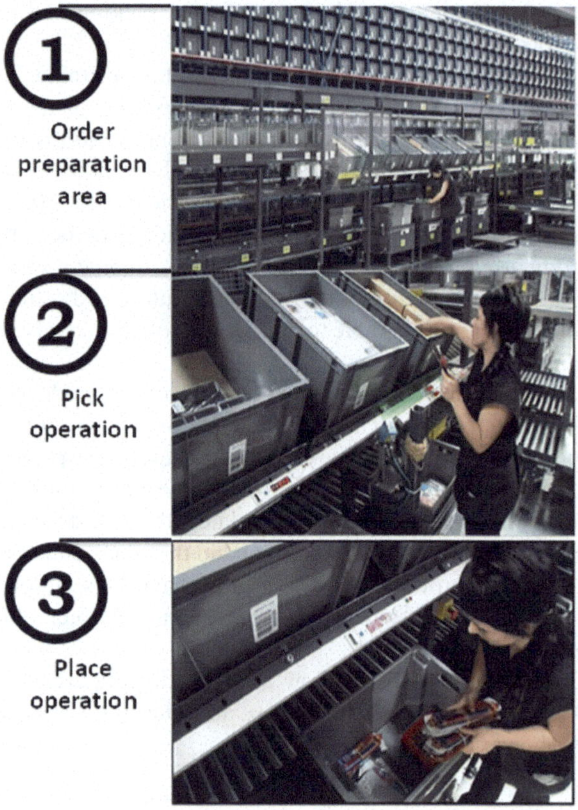

Fig. 6.4 Order preparation process in the last stage of an automated material handling solution

- Injuries reduction: despite ergonomic considerations during the design of layouts, order picking can result in a variety of muscle, joint injuries and pains from repeatedly or, forcefully reaching for and lifting product, using scanners, applying stickers, standing in one position or working overtime.

6.2.1 Requirements of the Solution

The high variability, location and positioning of items to be manipulated is represented in Fig. 6.5, which shows some physical features of the elements to be picked. The location of the elements in the box is also variable. The goods can come in diverse boxes and diversely ordered as shown in Table 3.2, and can have different heights (600 mm × 400 mm in base and 450, 320 and 235 mm height). The order in the target box as shown in Fig. 6.5, depends on the objects to be manipulated. When the product has an orthogonal shape (boxes), it has to be placed in an ordered way. While, when the products are difficult to be ordered the placement can be random.

Fig. 6.5 Requirements of items to be manipulated and random position in the containers

The Use Case technically combines two classical well-known general problems in one: bin picking and bin packing. In bin picking objects of different size and shape must be identified and grasped by a robot (mobile manipulator). In the bin packing problem, objects of different volumes must be packed into a finite number of bins or containers in a way that minimizes the number of bins used.

Specifically, the final project **technical objectives** are the following:

1. **Real-Time fused 2D-3D picking box segmentation**. For real-time perception, it is proposed the use of 3D point clouds combined with 2D colour imaging. The approach is based on processing images of a RGB-D camera. This enables to extracting information about the objects in the scene with a very low latency for further decision-making and planning stages. By combining both approaches, the number of descriptors available for each estimated object is much higher, allowing much more precise detection. In those situations in which the objects are closely packed or can occlude each other, 3D vision may have problems to detect their boundaries. When dealing with non textured objects if 2D vision cannot extract the needed features, 3D vision can be used to perform an object matching using shape information.
2. **Workspace monitoring for safe robot behaviour**. The objective is the development and evaluation of an environment monitoring system to detect humans around the robot and modify the robot's behaviour in order to meet the security

requirements for safe human-robot coexistence. The proposed system will detect and track humans around the robot, extracting information about the distance to the robot and their trajectory using laser measurement systems and RGB-D sensors for data acquisition, placed on the robot base in order to monitor all the workspace around it.

3. **Collaborative workflow with automatic mosaic design** for packing: In order to orderly pack all the elements (heterogeneous set of products) in a box, it will be necessary to develop a software able to automatically generate box mosaics from a heterogeneous set of products. Artificial Intelligence algorithms are combined to provide the best mosaic from a given combination of products that compose an order. The software will provide the mosaic to be performed by the robot and operator, and the sequence that has to be followed to retrieve the items from the storage racking to create the planned mosaic. In order to build a robust and flexible solution, the project has to deal with unexpected events during operation (item dropped from the gripper, inventory error causing empty product, human error, etc.) which imply the need of re-planning, a partial new mosaic configuration. This will be a significant functionality. Instead of using rigid task sequences, it is necessary to introduce flexibility in the task sequence providing re-planning mechanisms for both, robot and management system. Finally, FLECOOP solution must be integrated in the Warehouse Management System.
4. **Mobile manipulation for pick and place**: To allow the planning of trajectories in unstructured environments, sampling algorithms are proposed combined with different grasping strategies to ensure the manipulation of objects in dynamic workstations. The manipulation and navigation algorithms have to deal with uncertainties coming from the environment and identification systems.

6.3 Activities and Results

The objectives defined in the project address key technical and industrial challenges in growing complexity through the Free style, Show case and Field test experiments performed during the EUROC developments stages as it is shown in Fig. 6.6.

6.3.1 Free-Style: Fast Mono-Reference Pick and Place with High Availability

The Free-style demo develops a specific task of the final use case to be performed by the platform, where items are picked from a box (same reference) and placed in the final order box at a high cycle time. From the industrial point of view the whole pick and place time span is the main constraint to allow economically competitive applications. The automatic box transport system puts the boxes automatically in the

Fig. 6.6 Free style, show case and pilot experiments stages

picking area. The robot has to grasp the piece, move to the packing box, place the piece at its position and orientation, and return to the origin for the next cycle. The minimum goal is to do it in 7,5 s, i.e. a throughput capacity of 500 cycles per hour. Any figure above will help to offer a more attractive solution for the market. The free-style demo is focused on: (1) monoreference items identification and pose estimation in the picking box and (2) safe, collision-free and fast trajectory generation, part grasping and transferring to a packaging box where they are positioned. The set of objects considered is a subset of the items to be picked in the final solution.

6.3.2 Show Case: Safe Multireference Integrated with Current Pick to Light Solutions

The show case will provide a mobile manipulator picking items from a box (different references) and packing them in a predefined order in a pick to light structure, being sensitive to human presence around the robot. From and industrial point of view the requirements are to get a minimum cycle time with high availability. From human detection point of view, the objective is to reach a 100% of detection rate. The show case is focused on: (1) multireference identification and pose estimation in the picking box and (2) safe monitoring of the robot workspace. The variability, location and positioning of items considered. The system will detect and track humans around the robot, extracting information about the distance to the robot and their trajectory, to modify the robot's behaviour in order to meet the safety requirements. Picking box segmentation will be extended cope with multireference items in the same box. And mobile manipulation for pick and place, as operator will coexist with the robot at the same area the robot will be integrated with current **Pick To Light systems**, this way the operator will be helped by the system in the same way of his ordinary labour. Based on previously developed manipulation strategies the robotic navigation algorithms are adapted to position elements in the pick to light structure.

6.3.3 Field Test: Collaborative Solution for Unitary Picking in Order Preparation Area of a Distribution Center

The use case will be implemented at End User premises. It will provide a fully functional prototype of a hybrid automation solution integrated with the warehouse management system. Multiple references are picked by the robot and the operator from their corresponding boxes and then, packed orders are composed collaboratively. Robots autonomously pick objects and react safely to people position around them. Collaboration schemas are implemented for the proper mosaic building in the packaging box. At this stage the industrial objectives are more demanding, having in mind that a complete solution deployment at end user premises will be achieved with a TRL7. The Use Case focus on Collaborative workflow with automatic mosaic design for packing. The order preparation system will be coordinated with the FLECOOP system in order to coordinate robot and human activities. In this sense, some functionalities have to be developed: flexible mosaic building system for multireference items and collaborative human-robot workflow system; In addition, for some mosaic composition control force mechanisms have to be considered. Furthermore, the Use Case integrates previously addresses technical objectives in the end-user premises including: (1) Robust items recognition and pose estimation system; (2) A mobile manipulator with fast and safe trajectory execution, and pick and place capabilities.

6.3.4 Technical Approach

As stated in the introduction, unitary order preparation remains as one of the main challenges in Logistics. The FLECOOP Showcase approach has been built on top of the following premises coming from previous works [1–3], in warehouse solution development to build an autonomous unitary picking solution:

- To build a reliable solution with the combination of state-of-the-art approaches in perception and manipulation with industrial expertise.
- Heuristic approaches have been proven a more promising route to design perception systems rather than general purpose approaches.
- Hardware and software work in synergy to provide an efficient solution.
- The importance of error recovery mechanisms (dropped objects, destroyed objects, or miss-classified objects) for building a reliable system.
- To include Human in the loop strategies to increase systems availability.

The approaches in other works are different depending on the specific requirements of storage management solutions. **Amazon Picking Challenge** [4] focus on preparing orders by picking specific objects from shelves and placing those objects into specified containers (based on their goods-to-person automation, warehouse management solution). Similar approaches are the Scallog System™ [5] or Fetch Robotics [6].

The difference with this approach is that ULMA Handling warehouse solutions (as well as other warehouse solutions) are not based on this approach (having products on shelves), but they have big automated warehouses where the products are introduced in containers. But some challenges are common in the warehouse community such as:

- Identifying multiple objects placed on a variety of poses. The perception problem is an example of the general object detection, segmentation and poses estimation, which is widely researched in computer vision. Most of the approaches use 3D vision systems, with different segmentation strategies based on multiple features (colour, edge, height), foreground subtraction and color histogram, etc. [1, 3], or combining 3D and 2D features. In line with these approaches, FLECOOP combines 2D vision with 3D point clouds in order to maximize the identification rate.
- The manipulation in restricted spaces like boxes and shelves that leads to difficult high-dimensional motion planning problems. Magazino's approach [7] is based on ad-hoc mobile robots (TORU Cube) for picking individual objects in warehouse shelves. Similar approaches are also proposed by LocusRobotics [8]. The solution is limited to handling boxes that in many warehouse solutions can be enough but not for unitary picking. In this sense FLECOOP is more flexible and oriented to unitary picking.

Up to now, there are no unitary order preparation robotic solutions on the market, although multiple companies are looking for this kind of solutions as it has been seen in a recent visit to the Logimat 2017 fair were different proof-of concepts have been shown. Swisslog is developing an order picking solution called AutoPiQ [9] which objective is very similar to FLECOOP objectives, with several differences in the solution from the hardware configuration and approach point of view: it is focused to repeated single item picks, orders are not prepared following a mosaic of different products, and the perception system is integrated in the platform. The 3D and 2D approach seems quite similar but more detailed information is missing. On the other hand, although it is intended for allowing human to work alongside it not seems conceived as a collaborative solution as in FLECOOP, human and robot working in a coordinated way building a mosaic. Another interesting aspect is the grasping problem and the idea of building multipurpose grippers that are more suitable to deal with the problem of multiple references. There are other approaches that participated in several Amazon Picking Challenges (APC) such as DoraPicker [2] where different gripping approaches are analyzed.

6.3.4.1 Logistic Workflow Controlled by Warehouse Management System

The order preparation and product return solution has been integrated with the General Warehouse Management System of ULMA.

This system is in charge of coordinating all the elements in the end-of line of the Distribution Center, as it is described in Fig. 6.7, which is composed of the robotic units including sensors, the conveyors and the interaction elements with the operator.

The system workflow is represented in the right side of Fig. 6.8. The orders are defined by the WMS, and sent to the robot units. In an order preparation task, the WMS approaches the containers to the fixed robot in order to pick and place operations. In a good return task, the Miiwa picks the elements from a container and moves to the specific containers.

During the Field Test the main software components that control de robotic units have been developed. These are represented in Fig. 3. The communication between the Warehouse Management System (WMS) and the Robot Management System (RMS) is based on Web Services. The WMS will notify to the RMS module each order preparation or return task that has to be performed, and the RMS will send a confirmation as feedback after a product has been picked and released. In case that an error occurs, it will be notified after the complete order or devolution task is finished (Fig. 6.9).

The description of each of the components is as follows:

- **Vision Module**: This module is charge of the image acquisition of the 3D camera. Once an image is available, two services are activated; one for the bin location and

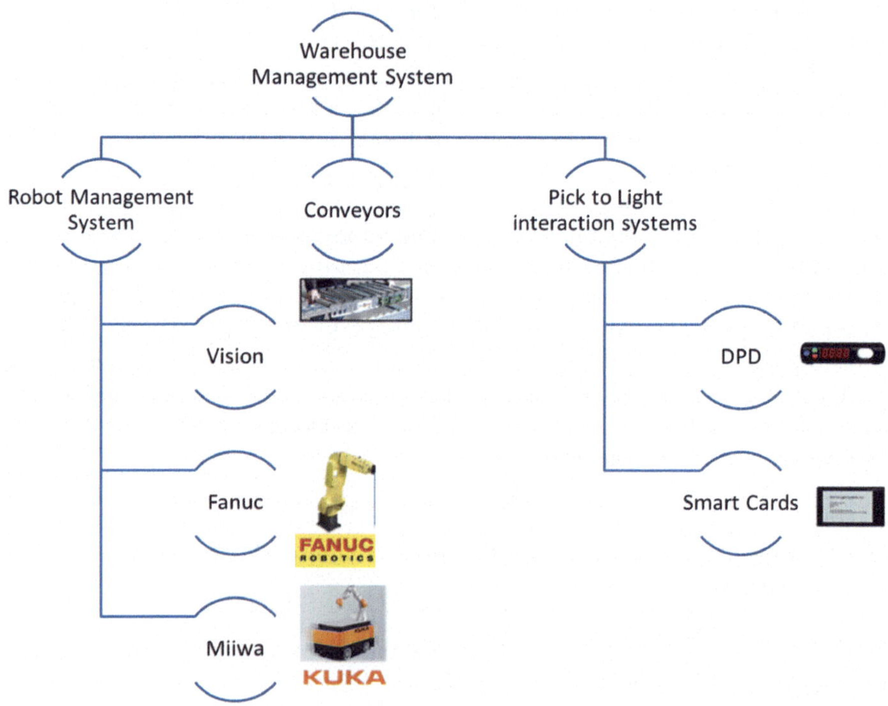

Fig. 6.7 Global architecture of WMS. Main elements

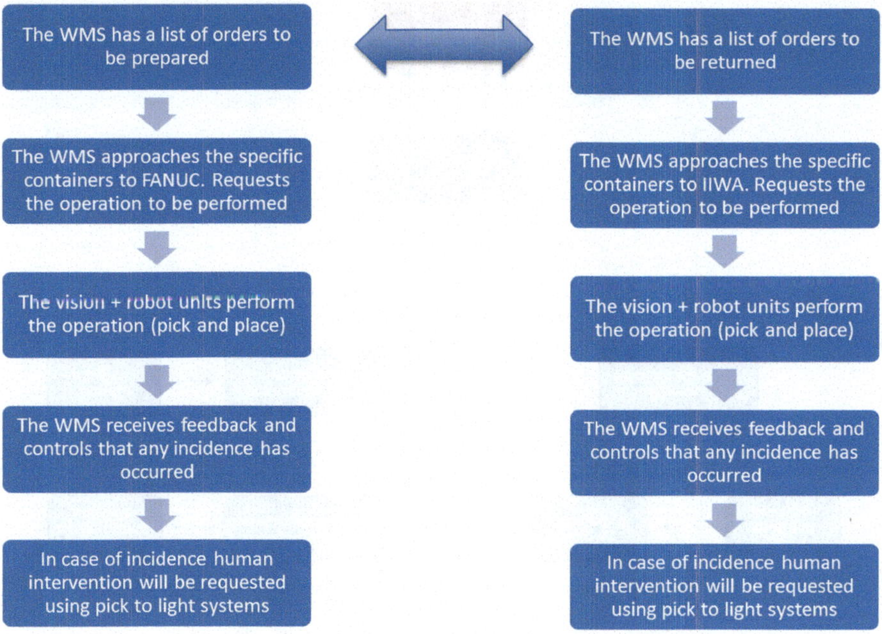

Fig. 6.8 WMS workflow for order preparation and return of goods operations

another one for the product pose estimation. The vision module returns the list of graspable products.

- **Pick/Place Module**: To ensure a complete successfully grasp and release process this module provides a collision free path planning and a monitoring of the product until it is released. It also includes the product identification functionality for return operations,
- **Mobile platform positioning**: This module implements an on-line calibration for Miiwa/KMR (as the pose of the robot is variable) to ensure the reachability of the products and to be able to know how the products in relation with the Miiwa base are.

6.3.4.2 Robust Item Detection

The main requirements from perception point of view are the development of a fast and robust product detection and identification vision system for random bin-picking in moving containers. The approach combines 3D and 2D vision for object pose detection, completing with 2D vision for object identification. Several product configurations are present in the containers used for pick and place operations: similar products in a container (order preparation) and mixed products in a container (good return). The workflow of the vision module is represented in Fig. 6.10.

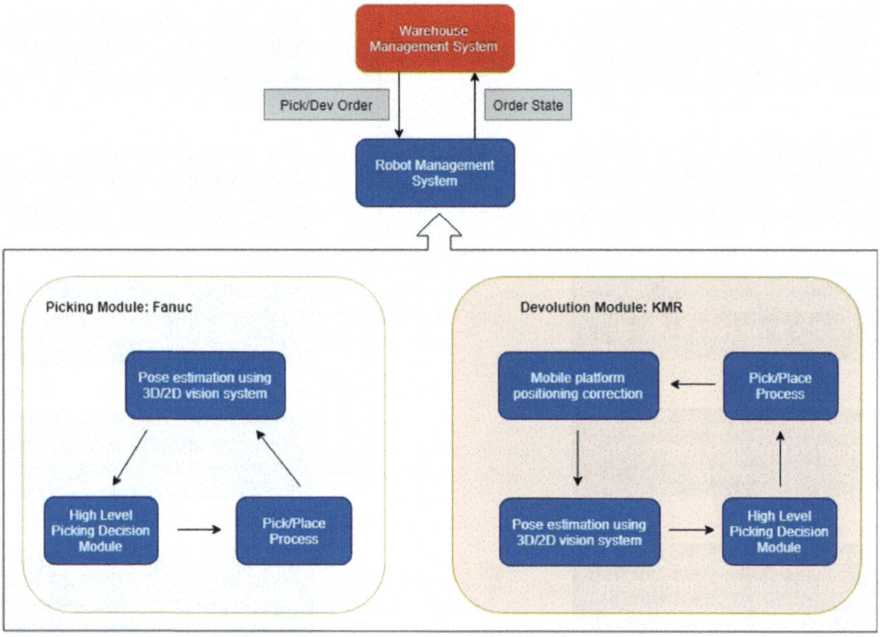

Fig. 6.9 Main software components that control the work flow for order preparation and return of goods operations

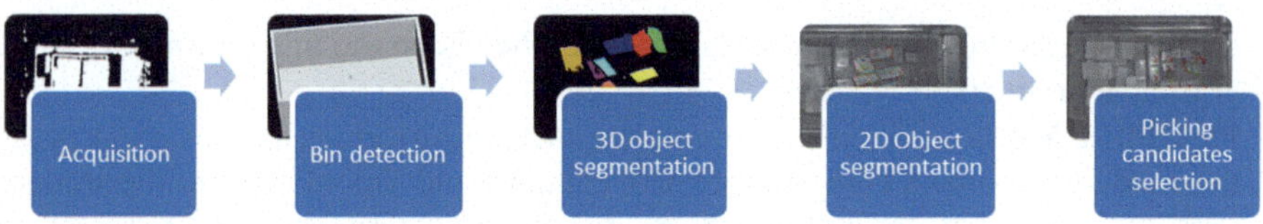

Fig. 6.10 Work flow of the vision module

The first step is acquiring a 3D scene, in this case using an Ensenso stereo camera, which provides a good balance between resolution and cycle time in this layout. This camera performs a calibrated stereo reconstruction using a noise pattern projection that helps solving stereo correspondence between image pairs acquired. The camera provides both a point cloud and a rectified 2D image of the scene.

Next step is to search the container/bin into the point cloud. A synthetic surface model of the bin is created starting from its dimensions (without CAD drawing). A surface model matching is applied obtaining the pose of the bin in camera frame, as it is shown in Fig. 6.5b. This pose is used (1) to reduce the initial point cloud focusing only in the region of interest inside the bin; (2) to obtain the location of the walls of the bin for the collision testing module. Locating the bin the system can adapt to any variability of bin in the conveyors.

Last, knowing the pick area inside the bin (this information is managed by the WMS) and the position of the bin itself, the resulting point cloud is analyzed. At this stage, as the elements are very close, it is usually difficult to discriminate different products only with the information of the point cloud. As a consequence, a 2D image (provided as well by the 3D sensor) is used, performing an edges extraction based algorithm, as it is represented in Fig. 6.6. As a result of the analysis, the vision module returns all the available planes found in the scene to the High level Picking decision module, which based on different criteria such us height, gripper orientation at grasping point, collision testing and robot trajectory planning selects the best candidate for grasping.

In good return task, the grasped product has to be identified in order to return it to its correct bin. To perform this identification task, a 2D camera is placed having a below perspective of the part. Here a scalable shape model (based on border information) is trained with relevant information of the faces of the different products. When the service is called an image is acquired and we try to match the product it appears with the trained possible models returning the best match achieved. At this point, it is worth to mention that considering the huge number of references to be manipulated in a ULMA installation, the solution provided for order manipulation works without any dimensional information or model, in order to simplify the configuration and maintenance requirements of the system. In addition, considering the throughput requirements mainly in the order preparation task, an optimization of the whole process has been performed achieving fast identification of products, in less than 1 sec.

6.3.4.3 Robust Pick and Place Without Collision. Manipulation Planning and Grasping

Two robots has been used to perform the order generation and returns tasks, specifically a Fanuc 200id 7 L in the picking process and KMR in return. Both robots are controlled under ROS middleware. The communication with Fanuc is established through a ROS Industrial based driver, while with the KMR the driver developed by DLR has been used. Both drivers have been designed to accept PTP (Point to Point) movements in joint space. However, the vision module provides bin location, as well as products pose in the Cartesian space, requiring a kinematic solver that provides the conversions to joints space and vice versa. Another factor to consider is the collisions that may arise when accessing into the container to pick the products. To implement collision free solution, it has been decided to use Moveit!. Initially, OMPL has been selected for planning the trajectories for pick and place operations. However, several problems arise:

1. Due to the high number of devices that are connected in the same network (two 3D cameras, one 2D camera, two robots (one of them with wireless connection), Warehouse Management System, Pick to Light), to achieve continuous movements and not bumpy trajectories, a reduction of the dimension paths is required.

The trajectory generated by OMPL planner contains a lot of points with the additional drawback of execution time (with higher cycle time). With the aim of reducing the path and the cycle time STOMP has been selected as a new planner for both planners.

2. Another disadvantage is that OMPL is probabilistic planner that doesn't ensure an optimum path planning. This problem is solved with STOMP, since it generates an optimum path based on the current robot state and the allowed standard deviation in positioning (the lower this parameter is, the closer the solution will be to the robot current state). In addition, in order to provide an optimal kinematic solution, instead of KDL, IKFast library has been used, which is a kinematic solver algorithm generated specifically for each robot (based on their own model).

To avoid collision risk with the container, once the bin has been detected with the vision module, it is added to the planning scene as it is shown in Fig. 6.11, considering all added elements as collision objects. After a path generation, using FCL (Flexible Collision Library) it is verified that all trajectory points are collision free in the current environment.

The cycle time is one of the most demanding requirements set by the end user. To this purpose, all the product identification analysis has been developed in a parallel thread, so once the product has been picked and the release process starts, a new identification analysis is launched, being possible to start immediately the picking of the next product just after finishing the previous one. The vision module provides all the products identified in one analysis. That list is processed by the High-level picking decision module, applying different kind of criteria, such as, height ordering, orientation restrictions and collision checking (only the tool and the bin are considered, it is independent from the arm), obtaining the best graspable products. This approach allows to build a more robust solution that in case of any problem with the first graspable product (e.g., no collision free path), a new attempt could be performed with the next product, without a new vision analysis and thus, improving the cycle time (Fig. 6.12).

The robustness requirement demands to build a system able to deal with potential incidences that may occur during an order preparation cycle. To this purpose several strategies have been implemented:

- Both suction grippers are endowed with a vacuum switch to monitor that the picking process has been executed successfully. In case of error during the picking, a second short linear movement is executed in Z axis tool in order to try to solve the error that could occur in the product pose estimation. If the second attempt fails, the grasping task will return an error, and the picking of the next detected product will be executed.
- If the picked product falls, the system is able to detect it and performs different strategies depending on the operation being performed (order preparation or return), in order to minimize the impact in the cycle time. In order preparation, two more attempts to pick the product are performed; in a return task, as a mobile platform is involved, for safety issues, the error is notified to the WMS.

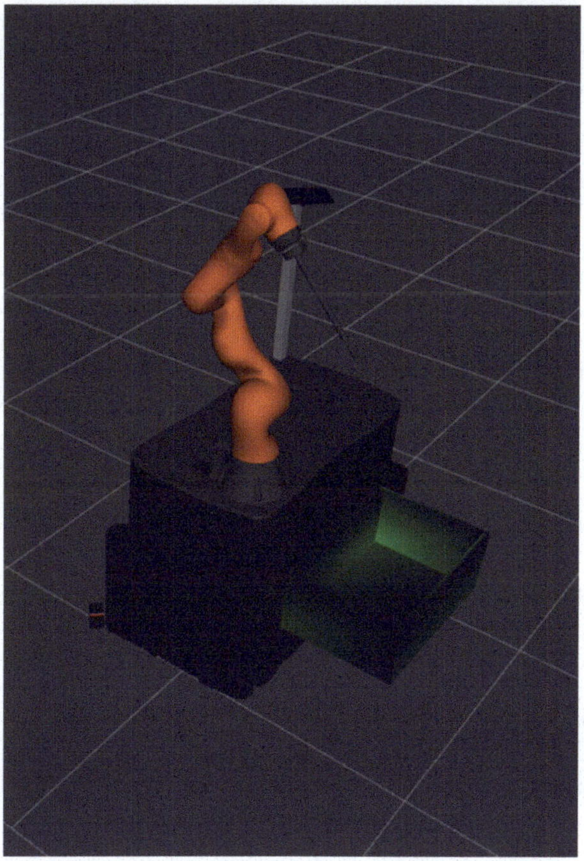

Fig. 6.11 Miiwa robot with added collision object

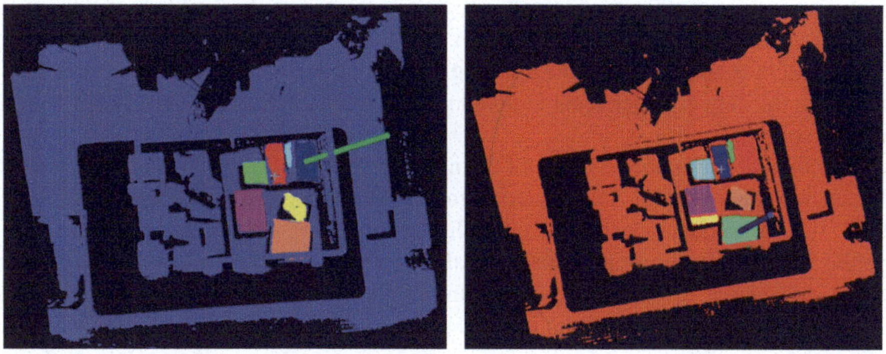

Fig. 6.12 Product with collision (left); A collision free grasping product (right)

- The path planning of all the grasping movements (composed by arm positioning in the middle of the bin, previous picking pose, grasping pose and the post picking pose) is done before the execution, just to ensure that all those poses are achievable for the robot and avoid the robot for being stuck inside the being because due to range problems.
- After a vision analysis, if none products are detected, another two attempts will be available before notifying to the WMS that something strange is happening in relation with the camera.
- During the product return task, if the product is not recognized, the tool will rotate several times trying to perform the identification successfully. In case that the identification fails, the product will be left in the original container and another grasping process starts. After two more attempts if the devolution is not finished properly, the WMS will be notified and the manual return with operator intervention will be activated.

6.3.4.4 People Detection and Tracking

The People Detection and Tracking module is in charge of monitoring the area surrounding the mobile platform. To this end, several concentric danger zones, centered in the platform, can be defined by the operator. The system is able to detect if a person enters those areas and to display information about it in a visual way as well as to send it as a ROS message to be read by other ROS nodes governing the platform or arm movement.

People detection is performed analyzing the lectures from a Hokuyo laser range scanner. These lectures are only in a plane perpendicular to the ground and the device is not mounted in a static place. People are detected as clusters of readings closer to the laser than some "ground truth" previously computed. This "ground truth" could be obtained directly from the environment or user-defined. The last case is particularly useful when we are interested in only a region of the scene observable by the laser. If there is a global map the mobile platform position with respect to the laser is known through ROS tf messages and the tracking system takes into account this information to update the danger zones localization and to assure that the platform is never taken as a person.

It is always maintained a count of the number of persons detected, as well as their distances to the platform. There is a designated area for people to leave the monitored zone. If a person disappears, due to occlusions or going out of the monitored zone by the wrong exit, the system activates the maximum alert. It maintains a counter of the number of disappeared people, and stays in maximum alert mode until all disappeared people appear again in some place in the monitored zone. The followed trajectories are shown in an interface, along with a pictorial representation of the layout, the danger zones, and the followed trajectories and the trajectory predictions.

Trajectory prediction is performed using long short-term memory (LSTM) [10] recurrent neural networks. LSTMs belong to the growing field of deep learning paradigms. RNNs are artificial neural networks in which connections between units form a directed cycle. Due to this architecture, recurrent neural networks possess an internal state that stores information about past inputs. This endows the recurrent networks with the ability of processing sequences of inputs and exhibits a dynamic temporal behavior in response to those sequences. Unfortunately, training RNNs to learn long-term dependencies by gradient-descent methods has proven to be difficult [11, 12]. LSTMs address this problem by introducing gates that control how much of the past and the current state has to get through to the next time step. To our knowledge LSTMs had not been used to perform trajectory prediction of people in an industrial environment using data from a 2D range laser.

6.3.4.5 Dynamic Mosaic Building

The selected mosaic building approach aims at dealing with fast for on the fly re-calculation. This allows providing a dynamic mosaic building approach, recalculating the mosaics coping with variability in the item positioned in the container. The Mosaic Construction module is in charge of fitting small boxes (products) inside a bigger box (container). This is similar to the bin fitting problem. In the bin fitting problem the goal is to find the minimum number of boxes for packing a given set of boxes. This problem is known to be NP-complete; meaning the time needed by the best known algorithms grows exponentially with the number of the boxes. To deal with this complexity, greedy algorithms for the bin fitting problem have been developed. One of them is the First-Fit Decreasing (FFD) algorithm. It can be implemented in O(n log n). It cannot guarantee optimal solutions, but it is assured that does not use more than 11/9 of the optimal number of bins [13]. This algorithm only takes into account the volume of the objects, not their shape.

The main goal was to make a fast algorithm. To do so, we have modified the FFD algorithm to take into account the geometric constraints of the objects, as well as the rotations allowed. It is also necessary that all the objects have a minimal support, in order not to hang in mid-air or not to be in equilibrium. A support of a minimum of a 70% of the area of the base for every object is enforced in any valid solution.

A graphical interface has been developed (see left 6.13), with two tabs, one for selecting the number of different objects of each type and another for selecting the dimensions of the objects and the container. This data could come from a database and be selected by a combo box or drop-down list control. In the Mosaic labeled tab the orientations permitted to the objects can be selected. Some industrial environment configurations could permit to rotate the objects in some plane, all of them or none of them at all.

The algorithm is very fast when dealing with a low number of items. For example, it is usual to take about 2 ms for 12 objects or 60 ms for 120 objects in an off-the-self personal computer. When more objects have been considered, it have taken 31.51 s in trying to fit 3000 objects inside a cubic container where not all the objects will fit, or 49.45 s for fitting all the objects in a cubic container. In the first case the algorithms successfully fits 1827 objects using a 94.17% of the volume of the container. These running times are fast considering the nature of the problem. The contribution has

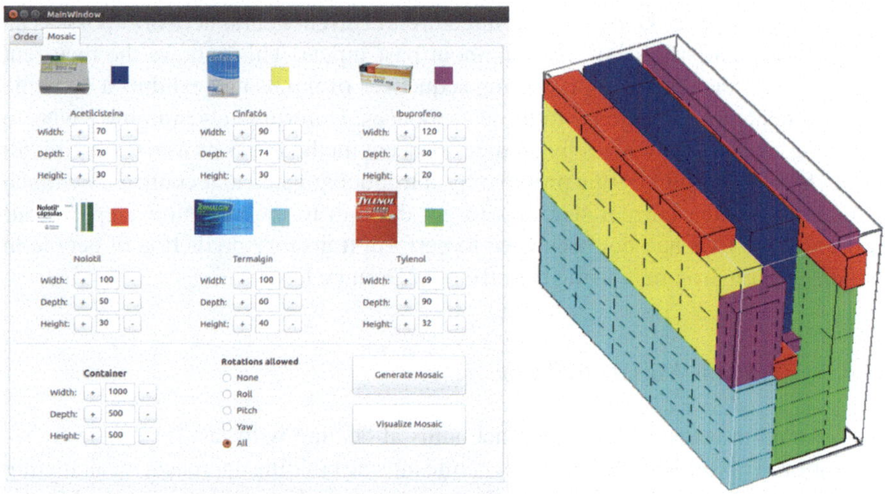

Fig. 6.13 Graphical interface for mosaic definition

been the adaptation of a well-known bin fitting algorithm to the geometric, rotational and support constraints typical of a real-world problem.

6.3.5 *Results*

The Field Test has provided a working prototype that works in a realistic environment as shown in 6.14. This prototype has moved to Logimat 2018 Fair, to show in the logistic market the potentiality of this solution. It is planned to have a real implementation at ULMA customer premises in less than one year. The system has been tested under different working conditions as demonstrated in the different experiment performed, and the prototype may be situated at TRL 7.

The specific experiments can be evaluated through the videos recorded (provided by the host and challengers) and the evaluation reports. All these experiments have shown the feasibility and the robustness of the solution, working in variable conditions in a semi-unstructured environment, using a mobile manipulator and introducing a small but representative number of references. Further experimentation and evaluation is required in a more integrated solution with the automated warehouse in order to obtain relevant KPI.

The technical results obtained lead to a complete prototype able to prepare orders and manage good returns. It is planned to implement the first customer solutions in a year after the EUROC project. The team members are developing their own exploitation strategy based on the achievements of FLECOOP, and mainly ULMA is implementing its strategy to reach the market with a unitary picking solution in

Fig. 6.14 Final prototype

medium term. The full system resulting will be directly deliver to the market by ULMA, while the others (hardware and software components/systems) will need to reach to an exploitation /collaboration agreement with different players within robotics players.

6.4 Conclusion

Current customers demand highly diverse set of orders, more and more customized and personalized products and shorter order lead times (the time between placing their orders and receiving the goods). This leads to the need of the development of systems in material handling and logistics industry capable of supporting this trends.

In this sense FLECOOP results impact technologically in:

- New hardware and software systems able to reduce errors and cycle time associated with manual picking and packing.
- Safety systems that allow to organize hybrid human-robot picking operations more effectively.
- Technological background for building the right balance between automation solution and manual processes to achieve high automation with high flexibility demands. Unlike their conventional robots working behind the fences, collaborative robots are lightweight and flexible, can easily be moved and reprogrammed to solve new tasks, meeting the short-run production challenge faced by companies.

FLECOOP objectives are aligned with ULMA Handling company strategy in medium term.

ULMA is leading the development of Unitary picking systems for material handling and logistics industry, with a wide market segments (pharmaceutical, consumer, etc.).

This innovative Unitary Picking System Solutions to be provided to the market by ULMA has to fulfill several requirements in terms of the products to be manipulated and the solution characteristics.

ULMA is dealing with current customers increasing demands of highly diverse set of orders, more and more customized and personalized products and shorter order lead times. Although the project has been focused on e-commerce products, similar features can be found in other product families to which similar solutions can be provided in case the experiment validates the results such as textile, pharmacy, ceramics, etc. In addition, although a complete mobile manipulation solution is considered in the project, partial developments of the complete solutions can be considered that can impact in different markets:

- Mosaic design system: 100% pallet order picking solutions in distribution centers (Food and Beverage, Pharmacy, etc).
- Robot with no collaborative abilities: 75% of pharmacy and bank field solutions (in these cases references (medicines, money) cannot be manipulated by people.
- Complete solution with collaborative mobile manipulators. 70% e-commerce field.

One of the main lessons learned among the partners of RSAII team is that the **complementary** between the participants and the tight work schema with clear deadlines allow one the one hand, to **accelerate** the project developments and on the other hand, to have a clear focus on project results achieving more **advanced prototypes**. This is aligned with the need of companies such as ULMA to provide new solutions to their target markets in **short term** since need to beat competitors that are developing research activities in this area. The complementary partners is also enriching, due to the different aspects to tackle, from more basic developments to final application, the full value chain must be represented. To achieve the FLECOOP Field Test objectives the work has been distributed according to the expertise of each of the partners. The plan has been performed without major deviations. The team have shown the required complementarity providing, on the one hand, the scientific background (UPV/EHU), the application oriented research capability (IK4-Tekniker) and on the other hand, the industrial expertise (ULMA Handling). This multidisciplinary team is composed of experts specialized both in robotics and smart and autonomous system development; they have been actively collaborating for many years (Fig. 6.15).

Complex bin-Picking systems have until now been developed generally using one of the two main different techniques, that is, 3D point clouds or calibrated 2D machine vision. The combination of 2D and 3D techniques in the same Bin Picking solution is an approach that has been hardly employed before in these industrial systems, being the main added value of this proposal. This information fusion approach will provide advantages in situations where objects are packaged tightly, with massive occlusions or with non textured objects. The processing and fusion of both approaches is computationally demanding, however, recent techniques manage to identify objects in the scene under Real-Time paradigms.

Fig. 6.15 RSAII team

References

1. Jonschkowski, R., Eppner, C., Höfer, S., Martín-Martín, R., Brock, O.: Probabilistic multi-class segmentation for the amazon picking challenge. In: 2016 IEEE/RSJ International Conference on Intelligent Robots and Systems (IROS), pp. 1–7 (Oct 2016)
2. Zhang, H., Long, P., Zhou, D., Qian, Z., Wang, Z., Wan, W., Chen, Y.: An autonomous picking system for general objects. In: 2016 IEEE International Conference on Automation Science and Engineering (CASE), pp. 721–726 (Aug 2016)
3. Eppner, C., Höfer, S., Jonschkowski, R., Martin-Martin, R., Sieverling, A., Wall, V., Brock, O.: Lessons from the amazon picking challenge: four aspects of building robotic systems. In: Proceedings of the Twenty-Sixth International Joint Conference on Artificial Intelligence, IJCAI-17, pp. 4831–4835 (2017)
4. Amazon robotics challenge.: https://www.amazonrobotics.com/#/roboticschallenge. Accessed 2017
5. Scallog system.: https://www.scallog.com/en/. Accessed 2017
6. Fetch robotics.: https://fetchrobotics.com/. Accessed 2017
7. Magazzino simple storage.: https://www.magazino.eu/?lang=en. Accessed 2017
8. Locusrobotics.: https://locusrobotics.com/. Accessed 2017
9. Swisslog autopiq.: https://www.swisslog.com/en-us/. Accessed 2017
10. Hochreiter, S., Schmidhuber, J.: Long short-term memory. Neural Comput. **9**(8), 1735–1780 (November 1997)
11. Hochreiter, S., Bengio, Y., Frasconi, P.: Gradient flow in recurrent nets: the difficulty of learning long-term dependencies. In: Kolen, J., Kremer, S. (eds.) Field Guide to Dynamical Recurrent Networks. IEEE Press (2001)
12. Bengio, Y., Simard, P., Frasconi, P.: Learning long-term dependencies with gradient descent is difficult. IEEE Trans. Neural Netw. **5**(2), 157–166 (March 1994)
13. Dósa, G.: The tight bound of first fit decreasing bin-packing algorithm is $ffd(i) \leq 11/9opt(i) + 6/9$. In: Proceedings of the First International Conference on Combinatorics, Algorithms, Probabilistic and Experimental Methodologies, ESCAPE'07, pp. 1–11. Springer, Berlin, Heidelberg (2007)

Chapter 7
TIMAIRIS: Autonomous Blank Feeding for Packaging Machines

Eurico Pedrosa, Gi Hyun Lim, Filipe Amaral, Artur Pereira, Bernardo Cunha, José Luís Azevedo, Paulo Dias, Ricardo Dias, Luís Paulo Reis, Nima Shafii, Andrea Tudico, Claudio Mazzotti, Marco Carricato, Simone Badini, Dario Rea and Nuno Lau

Abstract Current packaging machine vendors do not provide any automated mechanism for blank feeding and the state of the art is to have a human operator dedicated to feed the blank piles to the packaging machine. This is a tedious, repetitive and

E. Pedrosa · F. Amaral · A. Pereira · B. Cunha · J. L. Azevedo · P. Dias · R. Dias
Institute of Electronics and Informatics Engineering of Aveiro, University of Aveiro, Aveiro, Portugal
e-mail: efp@ua.pt

F. Amaral
e-mail: f.amaral@ua.pt

A. Pereira
e-mail: artur@ua.pt

B. Cunha
e-mail: mbc@det.ua.pt

J. L. Azevedo
e-mail: jla@ua.pt

P. Dias
e-mail: paulo.dias@ua.pt

R. Dias
e-mail: ricdays@gmail.com

G. H. Lim
School of Computer Science, University of Manchester, Manchester, UK
e-mail: hhmetal@gmail.com

L. P. Reis
University of Porto, Porto, Portugal
e-mail: lpreis@fe.up.pt

N. Shafii
NVIDIA, Munich, Germany
e-mail: nima.shafii@gmail.com

A. Tudico
University of Bologna, Bologna, Italy
e-mail: tudico.andrea@gmail.com

F. Caccavale et al. (eds.), *Bringing Innovative Robotic Technologies from Research Labs to Industrial End-users*, Springer Tracts in Advanced Robotics 136,
https://doi.org/10.1007/978-3-030-34507-5_7

tiring task. This also results in problems with unintentional errors, such as using the wrong pile of blanks. An alternative solution is the use of a fixed robotic arm surrounded by a protective cage. However, this solution is restricted to a single packaging machine, a unique type of blank shapes and does not cooperate with humans. TIMAIRIS is a joint effort between IMA S.p.A., Italy, (IMA) and the Universidade de Aveiro, Portugal, (UAVR), promoted by the European Robotics Challenges (EuRoC) project. Together, we propose a system based on a mobile manipulator for flexible, autonomous and collaborative blank feeding of packaging machines on industrial shop floor. The system provides a software architecture that allows a mobile robot to take high level decisions on how the task should be executed, which can depend on variables such as the number of packaging machines to feed and the rate of blank consumption at each one. Through a computer vision system, blanks of different shapes and sizes are correctly identified for adequate manipulation. The manipulation of the piles of blanks is performed using a single arm using compliant modes of operation to increase manipulation safety and robustness. Additionally, it has a safe navigation system that allows the robot to be integrated in an industrial environment where humans are present. Finally, it provides an enhanced multimodal interaction between human and robot that can be adapted to the environment and operator characteristics to make communication intuitive, redundant and safe.

Keywords Mobile manipulation · Planning · Industrial perception · Logistics

7.1 Introduction

The task of blank feeding a packaging machine, a known industrial logistics problem, is nowadays mostly performed by a human operator (see Fig. 7.1). It is a tedious, repetitive and tiring task for a human where occasional errors are unavoidable. On an effort to reallocate the human operator to more interesting tasks and decrease

C. Mazzotti · M. Carricato
University of Bologna - IMA S.p.A., Bologna, Italy
e-mail: MazzottiC@ima.it

M. Carricato
e-mail: marco.carricato@unibo.it

S. Badini · D. Rea
IMA S.p.A., Bologna, Italy
e-mail: BadiniS@ima.it

D. Rea
e-mail: ReaD@ima.it

N. Lau (✉)
Instituto de Engenharia Electrónica e Informática de Aveiro, Universidade de Aveiro, Campus Universitário de Santiago, 3810-193 Aveiro, Portugal
e-mail: nunolau@ua.pt

Fig. 7.1 State-of-the-art in the packaging industry. Blanks are feed manually (left image) or by a robotic arm in a cage (right image)

production downtimes due to mishandling of blanks, a fixed robotic arm surrounded by a protective cage was implemented by the industry as an alternative solution for blank feeding, as shown in Fig. 7.1. However, this solution is bounded to a single packaging machine, restricted to a unique type of blank shape and it is incapable to cooperate with humans (i.e. the robot and humans do not share the same physical space). This lack of flexibility is one of the obstacles that the INDUSTRY 4.0 initiative [1, 2] wants to remove from the industrial shop floor.

Research and implementation of flexible and autonomous solutions for logistic problems, such as blank feeding, can be accelerated by strengthening collaboration between the industry and the academia. This was the objective of the European Robotics Challenges (EuRoC) european project. By proposing a research project based on a robotic competition, it drives innovation in robotics applied to industrial manufacturing and logistics problems [3]. In summary, an industrial partner provides a use case and the academic partner offers a solution. This requires the exchange of knowledge between the partners to achieve a common goal, an innovative and productive solution based on flexible autonomous robots. The competition component of the project also has an important role; not only it promotes healthy competition between different partnerships but also offers additional funding for the most promising solutions. It was in the context of EuRoC that TIMARIS team was created. This partnership is a joint effort of the Intelligent Robotics and Systems (IRIS) group from Aveiro University (UAVR) in Portugal and IMA S.p.A. (IMA) from Italy. Together we proposed to develop a system for flexible, autonomous and collaborative blank feeding of packaging machines on industrial shop floor.

Our approach meets the requirements of a modern industry by rationalising the exploitation of raw materials, energy and repetitive human labour. The employment of robotised systems also increases the safety of workers' activity, since risky tasks can now be accomplished by automatised manipulators. This proposal intends to develop and transfer technologies from academia to industry to enable industrial robots with an architecture that allows them to: take high level decisions on how should the task should be executed (planning, replanning and optimization); integrate the robot in an industrial environment where humans are also present and moving (safe navigation; collaboration) [4]; be able to identify, locate and manipulate several types of blanks in an unstructured environment (computer vision, manipulation); benefit from the use of an enhanced multimodal human-robot interaction that can be adapted to the environment and operator characteristics, making it more intuitive, redundant and safe (communication).

7.2 Use Case Description and Motivation

Even in modern automatic machines, the final-product boxes are provided by the manual feeding of flat or pre-glued blanks. Loading blanks should be performed by robotised systems. This implies several advantages: it frees up the operator from the fixed rhythm, ensures an overall improved machine energy efficiency and leads to a more sustainable packaging process. Although the use of robots for manufacturing and logistics is common in the industry for several decades, the environment where these robots operate is, in general, heavily restricted, as is the case for most applications of robotic arms in industry, or the robots must follow a predefined set of paths that severely restrict their use in environments where humans are also moving. Interaction between the robot and humans is also quite limited and mostly performed by reprogramming the robot or by simple button-based input tools and LCD screens.

The TIMAIRIS blank feeding system intends to develop and transfer technologies from academia to industry to enable industrial robots with an architecture that: allows them to take high level decisions on how should the task be executed (planning, replanning and optimization); integrates the robot in an industrial environment where humans are also present and moving (safe navigation); is able to identify, position and manipulate several types of blanks in an unstructured environment (computer vision, manipulation); has an enhanced multimodal interaction between human and robot that can be adapted to the environment and operator characteristics to make it more intuitive, redundant and safe (communication). The end-goal of our team, within the EuRoC project (in which we participated in Challenge 2—Shop floor logistics and manipulation), was to develop a pilot of the "Autonomous Blank Feeding for Packaging Machines" task on an industrial shop-floor. But to reach the final goal, there was a set of preceding use-cases that had to be solved and evaluated as part of the EuRoC competition workflow. This workflow was an interesting innovation introduced by the EuRoC project. The use-cases that were developed before the Field Tests were the Benchmarking, Freestyle and Showcase introduced in the next sections.

The robotic platform used to address the use-cases, including the Field Tests "Autonomous Blank Feeding for Packaging Machines", was provided by the EuRoC Challenge host (German Aerospace Center: DLR—Deutsches Zentrum für Luft). It was a KUKA omnirob with a mounted Light-Weight-Robot (LBR) equipped with a two jaw-gripper and a stereo camera, and a pan-and-tilt unit also equipped with a stereo camera. Access to the cameras, the LBR and mobile platform was provided by a Robotic Operating System [5] (ROS) wrapper, hence all developments through the project were made using the ROS environment. This platform was common to all teams participating in EuRoC Challenge 2.

7.2.1 *Benchmarking*

In the Benchmarking use-case, all Challenge 2 participating teams had to solve the same two tasks defined by the challenge host. The first task was a logistics problem that required several small load carriers (SLC), which were distributed in the room, to be placed within a fixed target area. The objective was to recognize the SLCs using computer vision and use arm manipulation to pick them and place them in the target area as fast as possible without dropping the production materials inside the SLCs. The second task was a product pre assembly problem that required the use of a single anthropomorphic robotic arm to assemble several bolts with the proper washers and nuts with the help of a mounting fixture. The objective was to put the washers in the bolts and screw their nuts as fast as possible. The task of pre-assembling a single bolt requires several steps: identify the mounting fixture position: recognize the bolt, grab it and place it in the mounting fixture; identify a washer, grab it and place it in the nut; and finally find a nut, grab it and screw it in the bolt until the bolt's top is not occluded by the nut.

Without neglecting the competition part of this use-case, the Benchmarking was used by TIMAIRIS to learn on how to use the provided robotic platform and develop a software architecture to be used through the remaining use-cases, including the Field Tests. The end result is a skill-based anytime-agent architecture for logistics and manipulation tasks [6, 7] that was successfully used in all subsequent use-cases.

7.2.2 *Freestyle*

The Freestyle use-case was intended to develop and demonstrate technologies for collaborative problem solving between humans and robots using flexible multimodal communication. Based on the previous objectives, EuRoC teams were free to define the specification of the Freestyle task based on their expertise and on having in mind the final objective of their use case. The specification of the Freestyle task was performed in the submitted proposal to be admitted to EuRoC Stage II (benchmarking, freestyle and showcase). The TIMAIRIS task consisted in assembling a puzzle using a set of pentominoes pieces (formed by joining 5 cubes, face by face, on a single level). Assembling is assumed to be done cooperatively by a human and a robot. The highlights of TIMAIRIS Freestyle use-case are summarized in the following list:

- **Perception**: fusion of RGB/Depth information is used to detect the pose of colored and uncolored pentominoes, the pose of the fixture where the puzzle is assembled and also the puzzle state.
- **Navigation/manipulation**: planning and execution for safety/compliance in a human populated set. The robot has to pick pentominoes from the picking station, transport them, and deliver them in several different ways to the human, including placing it in the human hand, or assemble the pentaminoe in its final

position in the puzzle. Movements must be coordinated and safe, as the human also picks and transports pentominoes and assembles the puzzle.
- **Planning and replanning**: the proposed game can be very challenging for a human player. The high computational power of the robot allows the implementation of effective planning algorithms for solving the puzzle that can be interactively used to assist the human in finding the puzzle solution.
- **Multimodal interaction**: having several communication modalities that can be combined together allows for robot adaptation to different users and environments. During the free-style the following modalities were used: gestures; voice; graphical user interface (images and keyboard/mouse).

The Freestyle use-case was deliberately decided to be significantly different from the Showcase and Field Tests use-cases, but it demonstrates common functionalities with the showcase and field tests such as perception, navigation/manipulation, planning, multimodal interaction and cooperation.

7.2.3 Showcase

The Showcase use-case was composed of a simplified version of the complete autonomous blank feeding problem, that included a prototype of a blank magazine, the mechanism that feeds the blanks into the packaging machine, and two pallets filled in with actual blank piles used in production. The task consisted in recognizing the blank piles on two different pallets, devise a plan for picking them, and transport and place them in the blank magazine in a smooth way so it does not get jammed. The workspace is assumed to be shared by humans and the robot. The main issues addressed by this use-case are the following:

- **Perception**: Fusion of RGB/Depth information is used to detect the pose of the pallets, blank piles and also the pose of the blank magazine; multiple laser range-finders are merged to detect and track humans in the workspace. Compliance aplied to the manipulator movements is used to increase safety and to detect abnormal situations that result in non valid arm positions and also allowing the trigger of recovery procedures.
- **Navigation/manipulation**: Using a single arm, the robot has to pick blank piles from two pallets that require navigation planning and arm motion planning for picking and placing the blank piles. It should be noted that blank piles are composed of several distinct objects and have no obvious grasping position when laying on top of the pallets. For safety concern, the presence of a human in the vicinities of the robot triggers a generalized motion stall.
- **Planning**: Due to the blank piles arrangement on the pallets, there are piles that need to be picked first or, otherwise, manipulation will not be possible for all cases. Hence, that requires the generation of a plan establishing the picking order. Task execution is monitored while being executed and if deviations are detected

recovery plans are triggered. Detection of situations like "need for blanks" from the magazine is also included.

- **Multimodal interaction**: Having several communication modalities that can be combined together allows for robot adaptation to different users and environments. During the Showcase the following modalities were used: gestures; text-to-speech; visual feedback on a tablet; graphical user interface (images and keyboard/mouse). Humans near the robot are tracked using laser scanners and, when within the interaction distance, the pan-tilt camera keeps its attention on the closest human to recognize possible gesture commands.

The showcase scenario has been considered to solve pending problems for autonomous packaging. In automatic packaging, blank piles have to be manually fed by human operators, who may occasionally refrain from collecting blank piles from more distant pallets.

7.2.4 Field Tests

Since IMA produces and delivers packaging machines for many kinds of products such as pharmaceuticals, cosmetics, tea, coffee, dairy and food, and those products continue to rapidly evolve, human operators and autonomous robots are required to adapt to changing environments in a short period of time. This environment is less structured and demands higher flexibility than large-scale or mass-production industries. The Field Tests use-case aimed to demonstrate a solution for the blank feeding problem in the shop floor environment and using actual functional packaging machines, including sharing the space with human operators. There is a high motivation for automating labour intensive problems such as loading blanks. This is an essential operation, but one that involves little added value. The main ideas for enriching the final task compared with the Showcase (that was a simplified version from this final use-case) are mainly the following:

- **Blank identification and pose estimation**: capability for identifying several types of blanks as well as their orientation within the pallet, which may change from pile to pile;
- **Multiple machine feeding**: capability to feed more than one machine with the appropriate blank type;
- **Planning**: prediction and intelligent scheduling of blank feeding operations;
- **Pile blank grasping**: gripper with capabilities for grasping distinct types of blanks, the robotic solution also increases the human picking capability by allowing piles as high as around 15 cm and weighting around 6 kg to be pick in a single movement;
- **Safety management**: navigation and manipulation in an environment that is shared with human operators;
- **Advanced communication/interaction**: capability to interact with human operators through a multimodal interface.

The Field Tests setup required two packaging machines, and a set of pallets with at least two types of blanks (available at the end user). The position of the pallets and machines was similar to those found in real deployed shop floors. The orientation and position of the blanks on the starting pallet may vary according to the printing layout. Equipment also includes the KUKA platform and a new blank gripper, different from the one used in the Showcase. During these operations, the interaction between human, machine and robot clearly has to take place in conditions of safety, considering that the operator's job is to oversee the functioning of the machine and to intervene if anything goes wrong. The multimodal and bi-directional interaction between humans and the robot allowed an effective human machine interface (HMI) for robot-operator communication. Blanks are supplied on pallets, using flat blanks that are already creased and die cut. The pallets can be raised from the floor level to allow the manipulator mounted on the KUKA omniRob to grasp any block of blanks within the pile.

7.3 Activities and Results

7.3.1 Benchmarking

The goal of our team was to solve the Benchmarking use-case, proposed by EuRoC, with high precision, accuracy, robustness and as fast as possible. Time is a key factor while solving the tasks at hand. Not only it is an important evaluation metric, but for industrial application it is a matter of productivity.

7.3.1.1 Skill-Base Anytime Agent Architecture

The developed software architecture is divided in several functional components, as shown in Fig. 7.2. The *Perception* modules are responsible for collecting sensory data and process them to extract the relevant information used in high-level modules such as the *Solver*. A manipulation/navigation *Skill* is the capability of doing a particular action/task, such as, picking and placing an object or moving the manipulator's end-effector to a desired pose. *Perception* modules and *Skills* obtain and send sensory-motor data via the *Sensor* and *Effector* interfaces, respectively. It is the responsibility of the *Solver* to take decisions on how to solve the current task based on the available sensory information and available *Skills*. The *Action Planner* provides a sequence of skills in order to manipulate the current object of interest. Finally, the Order Graph represents the objects in the environment and the order in which they are manipulated. The order is restricted by a directed acyclic graph that represent a dependency between objects in terms of order of manipulation. For example, leaf object have to be handled before its parents, a situation that is ubiquitous in assembly problems (e.g. bolt assembly, puzzle assembly).

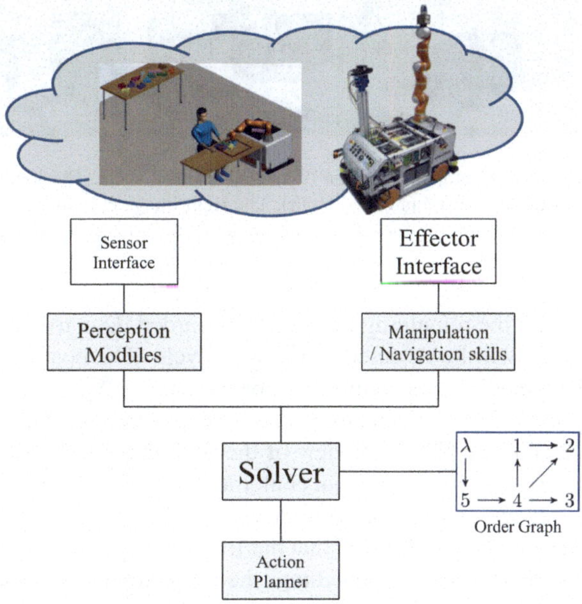

Fig. 7.2 Overview of the skill-based agent architecture

7.3.1.2 Production Logistics Task

The objective of this task is to find and pick all SLCs present in the room and place them in the designated area. Their approximate locations are known, i.e. on the table or in the shelf, but their exact location is not known. To guarantee that all SLCs are eventually detected the task starts with several search poses that cover the complete known area. Once a SLC is sighted its pose should be detected. Once the pose is obtained the agent decides how to pick and where to place it.

The SLC detection has two stages: first, color segmentation is used to candidate image regions that may contain an SLC; second, a 3D point cloud for each region is created and matched against a 3D template of the SLC to find its position and orientation (i.e. its pose). An example of SLC detection is shown in Fig. 7.3. The color segmentation, in HSV color space, takes advantage of the color homogeneity of the SLC and good contrast with the environment to reduce the search space. The output of this segmentation is a set of blobs that represent SLC candidates. However, blobs must have a minimum area to be valid. Overlapping SLCs in the image is a possibility, but that is not a concern as disambiguation is deferred to the next stage.

After color segmentation, the resulting blobs are used as masks to generate a pointcloud for each candidate. Let P be a pointcloud and p_i the set of points of P. To reduce the computational complexity of processing the pointcloud with a high number of points, P is decimated by applying a voxelization filter. Then, to remove possible outliers a statistical outlier removal is also used [8]. To handle the existence of overlapping SLCs, the point cloud is divided into clusters using an Euclidean clustering method [9]. The cluster with the highest number of points is assigned to

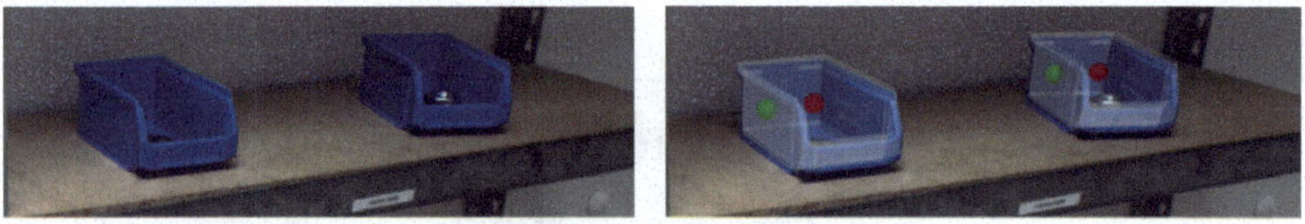

Fig. 7.3 Example detection of two SLCs in a shelf. The left image is what is perceived by the stereo camera (image plus depth). The right image contains a superimposed 3D model of the SLC for each detection, including their center (red dot) and possible pick point (green dot)

P and the remaining points are discarded. Due to our agent architecture, discarded SLCs will be detected in the next cycle. The position of the SLC is approximated by the pointcloud centroid c given by $c = \frac{1}{n}\sum_{i=1}^{n} p_i$ while its orientation is initially provided by the Principal Components Analysis (PCA) of the projection of P in the XOY plane [10]. The view of the SLC may provide a partial pointcloud that skews the centroid from the real center and the orientation of the PCA has an ambiguity of π radians. Therefore, the final pose of the SLC is calculated by matching P against a 3D template of the SLC that has its centroid in the origin and orientation aligned with the X axis. Before matching, the pointcloud P is transformed to its origin (i.e. its inverse pose applied to itself) so the transformation that results from the matching is the correction of the initial pose calculation. Because the orientation given by PCA is ambiguous, both orientations are used for matching. The pose with the lowest error is then selected as the final pose. The algorithm used for matching is the extension of the scan matching algorithm proposed by Pedrosa et al. [11] to three dimensions.

7.3.1.3 Product Assembly Task

The purpose of this task is to pre-assemble a set of bolts, nuts and washers using a single anthropomorphic robotic arm mounted on a mobile robot. To help the assembly, two fixtures are provided. The first hold the washers in (almost) upright position to help manipulation. The second has a well with the shape of the bolt's head, shallow enough to hold the bolt in place while the nut is being put in place. The bolts and nuts are in the vicinities of the fixtures (see Fig. 7.4). The approximate location of the assembly pieces and fixtures are known, but not their exact position.

The TIMAIRIS solution to this problem first moves the robot from the center of the room (mandatory robot location for starting this task) to a location near the table where the bolts, nuts, washers and fixtures are. Then, using the camera attached to the end-effector, the scene is observed from a point of view that provides a complete view of all necessary elements for the task without further need to change the position of the mobile platform. By keeping the mobile platform fixed, the relative position of objects to the manipulator is maintained and we can take advantage of this assumption to reduce the number of observations with the camera.

To execute the task the agent has to detect the bolt, pick it and place it in the bolt fixture, then detect the washer, pick it and place it in the bolt, and finally it has to

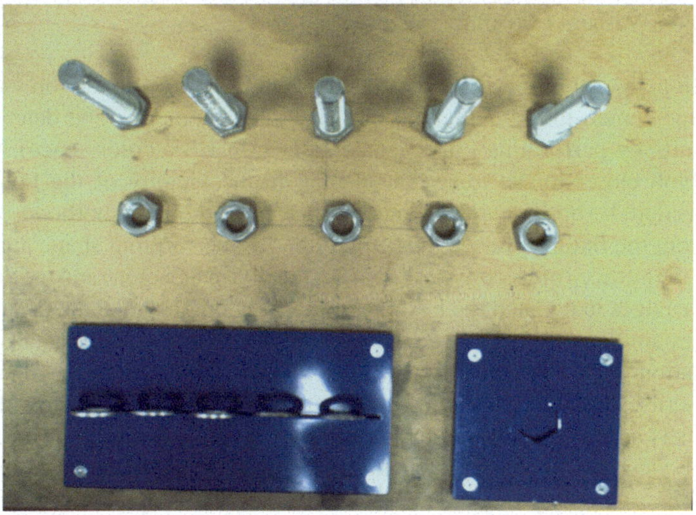

Fig. 7.4 Example of an initial setup for the product assembly task

detect a nut, pick it and screw it in the bolt. To calculate the positions of this objects we use the pinhole model instead of the pointcloud provided by the stereo camera. We can follow this strategy because the dimensions of all elements in the task are known and the distance from the camera to the working table is known in advance. By avoiding the creation of depth information we reduce the computational complexity and speedup the detection operation.

Fixture Detection

The detection of the fixtures, although executed only once, is an important step. Because the mobile robot has to navigate to the working area at the beginning of the task localization, some uncertainty in its absolute position is inevitable. However, the assembly area is well defined, therefore we can use the location of the fixtures to define the reference frame for the remaining elements. The detection of the fixtures is performed using color segmentation on the HSV color space. An initial, rough observation, is performed from the observation point of view. The resulting blobs are then used to calculate the rotated rectangle that best encloses each individual blob. The detection from the observation point does not provide enough accuracy for using the fixtures, therefore, using the information from the rotated rectangles, we approximate the camera to each fixture and repeat the detection individually for each fixture. The localization of the fixture that contains the washers is used in the detection of the washers and the center of the rotated rectangle that derives from the bolt's fixture coincides with the screwing well.

Bolt, Nut and Washers Detection

The detection algorithm for the bolts and nuts is the same. From the top view, close to the image center, these elements share a similar geometry. The detection of a bolt/nut starts by performing color segmentation (in HSV color space). Only the resulting blob closer to the image center and in the vicinity of the bolts virtual line is considered—both nuts and bolts are placed in the neighborhood of a fixed distance from the fixtures, hence the virtual line. By performing shape analysis we can obtain the vertices of its convex shape (i.e. ideally an hexagon) and try to find two consecutive edges that match the hexagonal shape in length and angle (see Fig. 7.5). These edges are then used to calculate the center of the bolt/nut and its orientation. The orientation is only needed for the nuts so that they can be picked by their edges.

The described approach assumes that the object to be detected is near the image center. This is achieved by generating a hint list for bolts and nuts, from the initial observation view using color segmentation, that provides an approximation of their position. When it is time to detect a bolt/nut, the hint is used to approximate the camera to the object of interest. The washer detection uses a different strategy. Instead of color segmentation an intensity filter is used to obtain the pixels with higher intensity (see Fig. 7.6). The resulting blobs are validated by their size. The centers of theses blobs are then used to calculate the position of the washers.

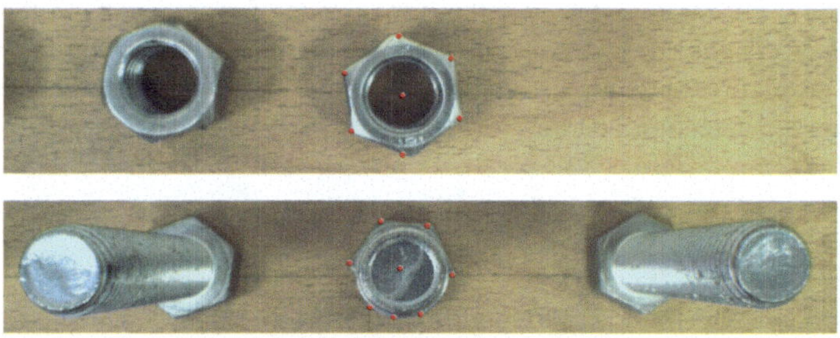

Fig. 7.5 Example of nut and bolt detection. The top image contains the detection of a nut and the bottom image contains the detection of a bolt. The red dots identify the object center and the shape vertices

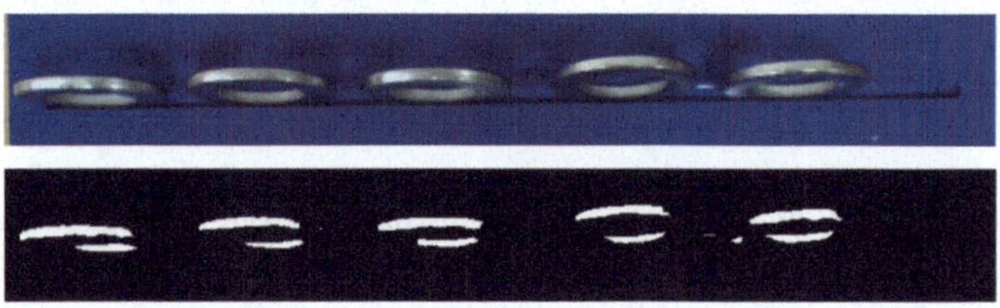

Fig. 7.6 Washer detection in the fixture. The top image is what is perceived by the camera. The bottom image is the result of the intensity filter

7.3.1.4 Motion and Manipulation

Extracting the information from the environment is only part of the task. Physical objects have to be moved in order to fulfill the task. Time is a critical metric in industrial applications. During development velocities were kept to a minimum to prevent possible physical harm and velocities were only increased when the combination of perception + manipulation proved to be reliable. In the Production Logistics task the base of the mobile platform was used as a buffer to store SLCs so that the number of motions of the mobile platform was optimized. In the Product Assembly, manipulation with compliance was fundamental. Using compliance, it was possible to put the bolt in the screwing fixture without visual confirmation by pushing and rotating the bolt until it fell inside the fixture well. Additionally, by using compliance, we were able to produce a soft vibration effect to the bolt while putting the washer in place that helped with possible stuck situations.

7.3.1.5 Results

The developed software architecture provided a solid base from which different industrial applications can be developed. Beside the two tasks of the Benchmarking use-case, all the remaining tasks of the EuRoC project were solved with the proposed architecture. It provided a high reusability of several components for different applications, most notably the motion and manipulation Skills. The perception modules that we developed for the Benchmarking use-case proved to be accurate, precise and had a high level of repeatability. The Production Logistics task was solved in 3 m:4 s while the Product Assembly task was solved in 6 m:15 s. For both tasks our Team, TIMAIRIS, obtained the fastest results among all EuRoC Challenge 2 teams with an advantage of 1 m:58 s in the Logistics task and 8 m:45 s in the Assembly task to the closest competitor.

7.3.2 *Freestyle*

The Freestyle use-case was performed using an area of around $16\,\text{m}^2$ ($4\,\text{m} \times 4\,\text{m}$). Setup consisted of a picking station, where the pentominoes initially lie, an assembling station, where the puzzle is assembled (the picking and assembling stations were not simultaneously accessible by the robot from the same spot), a set of 12 pentominoes (one of each different kind), a human player and the KUKA mobile platform (see Fig. 7.7). A depth sensor (e.g. Kinect) was used to recognize the gesture commands and two external monitors were used to convey feedback to the human/evaluators. The assembling station included a clearly defined area where the fixture that defines the position of the puzzle was placed by someone from the EuRoC C2 host. The picking station also had a clearly defined area in which the pentominoes were placed randomly by someone from the EuRoC C2 host. In those cases

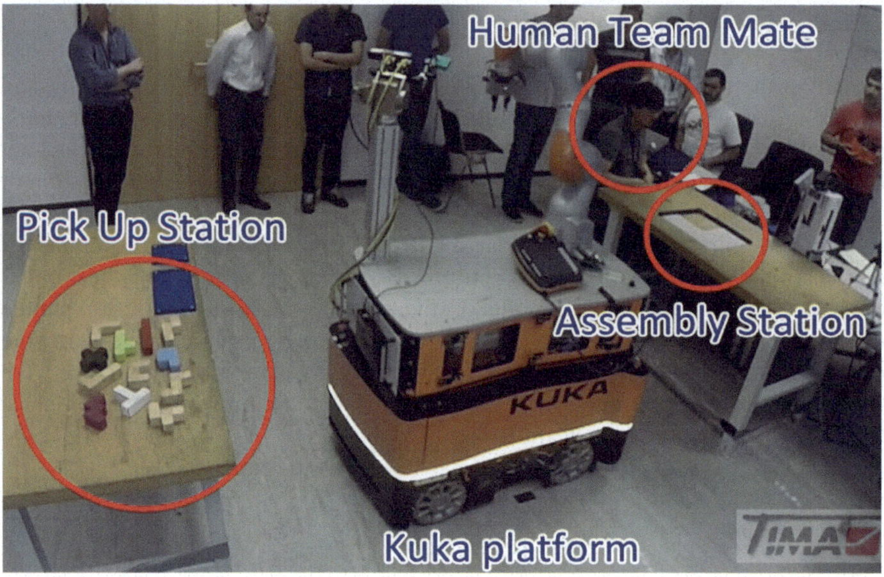

Fig. 7.7 Freestyle environment layout

in which not all pentominoes were used (challenges 2 and 3, see below), half of the pentominoes were chosen by a host member and the other half by TIMAIRIS members.

7.3.2.1 Pentominoes Puzzle Assembly

The task of assembling the puzzle is shared by a human operator and the robot. It is the responsibility of the robot (or robotic agent) to recognize the available pieces, devise a plan in order to assemble the puzzle and transmit that information to the operator (if requested). During assembly, the robot may have to replan, i.e. calculate a new puzzle solution, due to actions performed by the operator. Any action performed by the robot is always the result of a request by the operator.

Pentominoes Detection

The detection of the pentominoes is achieved by extracting shape information from an image combined with depth information (represented as a point cloud). Shape information is used because the pentominoes do not have unique colors and simpler approaches such as color segmentation failed to uniquely identify each piece. The first step of pentominoes detection is to remove from the point cloud the points that belong to the picking station tabletop. RANSAC is used to detect he tabletop points and all

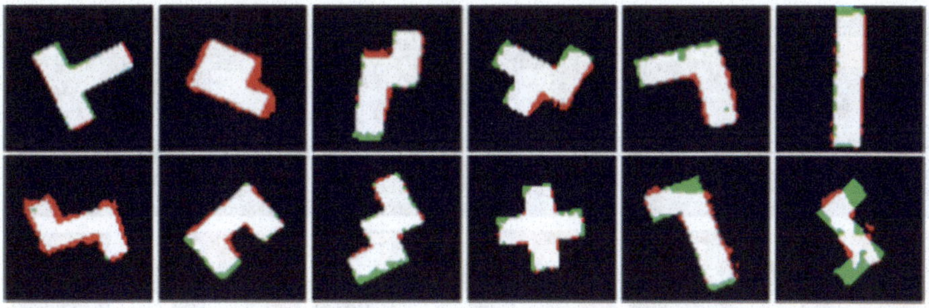

Fig. 7.8 Example of pentominoes detection with template matching. Each image corresponds to a detected pentomino that best matches the cluster of projected points

points below the tabletop (plus tolerance) and higher than tabletop plus pentomino height (plus tolerance) are removed from the pointcloud. The remaining points are assumed to be part of the top surface of the pentominoes. Those points are projected to the XY plane and then an Euclidean Clustering is used to extract individual pieces into clusters. Each of the clusters is then checked against all scaled templates of pentomino shapes in every rotation. Two metrics are extracted: percentage (%) of cluster area covered by the pentomino template, and % of template area covered by the cluster. The templates used for the 2 metrics are slightly different. The first metric uses pentomino templates that are slightly larger than the ideal pentomino template, so that small deviations of the cluster from the template do not decrease this metric, The second metric uses templates that are slightly smaller than the ideal template, to allow small imperfections on the border of the cluster. If the thresholds for both metrics are surpassed, then the pentomino position and orientation are discovered. This procedure allowed for a robust perception (e.g. Fig. 7.8).

Planning

Considering the available pentominoes, an assembly plan is computed considering the constraint that each piece has to be placed on the table and then pushed into its position. This constraint is what allowed us to also use the robotic arm (not just the operator) to assemble the puzzle. After each piece is placed in the puzzle the robot re-evaluates its current plan by comparing the current state of the puzzle with its internal representation. If both states do not match, it replans the puzzle assembly. The replanning situation happens when the operator does not follow the assembly instructions provided by the robot. An example of a plan is depicted in Fig. 7.9.

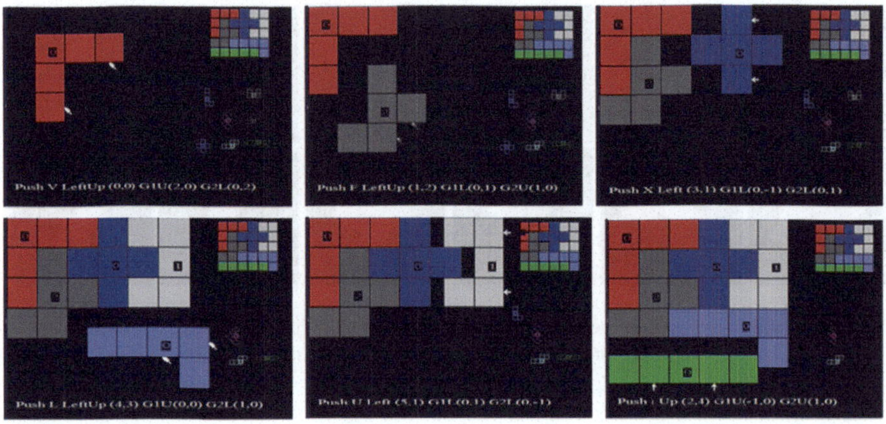

Fig. 7.9 Example of a puzzle assembly execution plan. Given the available pentominoes the plan provides the picking order and the push points and direction

7.3.2.2 Human-Robot Interaction Through Gestures

For the Freestyle use-case we introduced a Human-Robot Interaction (HRI) system to allow a human to communicate with the robotic agent through gestures [12]. This HRI was used to solve the pentominoes assembly in collaboration with the robot. A gesture language was used instead of speech to allow its usage in noisy environments, such as industrial shop floors [13]. This language is composed by eight one-hand gestures that form the base symbols, as shown in Fig. 7.10. Similar to human language, a limited number of symbols can be used to compose an unlimited number of sentences. For this use case we defined an interaction graph (see Fig. 7.11) that generates 24 sentences (or commands) using the available gestures. A sentences is composed by concatenating the gestures as sentences.

The gesture recognition is achieved with the combination of convolutional neural networks (CNN) and a contour based hand feature. The HRI system needs to construct a sentence from the continuous gesture recognition results. A confidence-based method is applied to concatenate a sequence of symbols in a robust manner [14]. An interval-counter for each gesture is defined on the basis of confidence law of inertia, where a recognized symbol is assumed to persist unless there is a confidence to believe otherwise. Additionally, applying a spelling correction from the word corpus [15], the HRI system can infer the complete command set with increased robustness. A complete description of the gestures representation and recognition can be found in the work presented by Lim et al. [12].

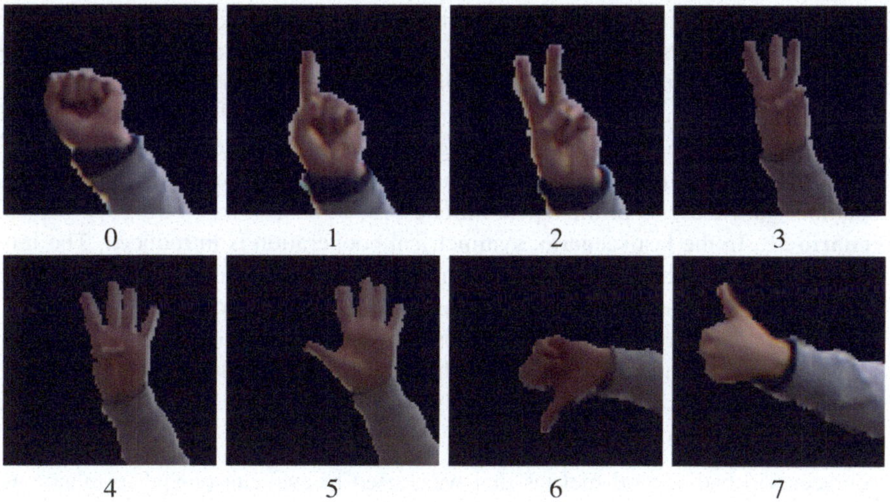

Fig. 7.10 The eight one-hand gestures and their corresponding IDs

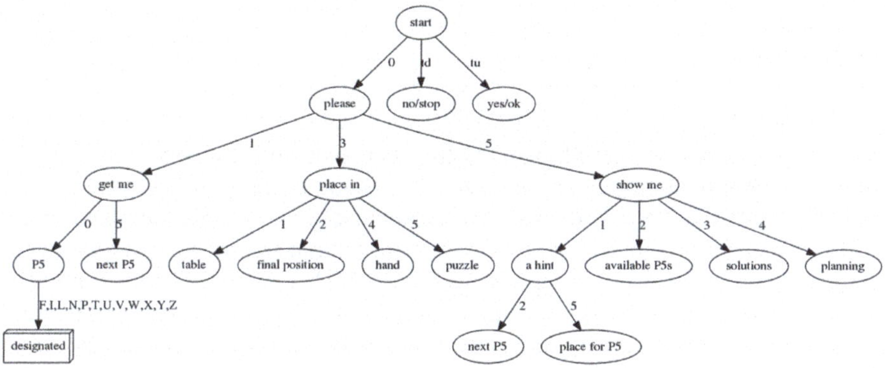

Fig. 7.11 Interaction graph that defines the communication with the robotic agent. Each leaf in the interaction graph correspond to an action performed by the robotic agent

7.3.2.3 Scenarios

The Freestyle use-case execution consisted of three different scenarios with increasing complexity. Here is their description:

Scenario 1 The human operator stays seated at the assembly station; he asks the robot to pick the next pentominoes for a possible solution; the robot gets it from the picking station; then the robot shows the final position of the pentominoes; the robot delivers the pentominoes; the human places the pentaminoe in the designated place; the process is repeated until the puzzle is finished; 12 pentominoes are available at picking station; the puzzle can consist of 5 or 6 pentominoes;

Scenario 2 The difference here is that the human always asks the robot to pick a specified pentaminoe; whenever needed, the human can ask the robot for a hint regarding the next move, which he can choose to follow or not; giving a hint implies that the robot must perceive the current status of the assembly process and plan or replan a solution accordingly; the robot can provide the number of solutions and the complete assembly plan; 8 pentominoes are available at picking station; puzzle uses 5 of the 8 pentominoes;

Scenario 3 In the last scenario, symmetrical cooperation is introduced. The initiative regarding the next move is given alternately to the robot and the human. Considering the characteristics of the task, both the robot and the human may ask the partner to fetch a pentaminoe or to assemble a given pentominoes in the puzzle; the robot can provide the number of solutions and the complete assembly plan; 8 pentominoes are available at the picking station; the puzzle to be assembled can consist of 5 or 6 pentominoes.

Each scenario had several metrics that were used to evaluate our performance as EuRoC challengers. Those metrics were focused on capabilities such as properly identifying the pentominoes and recognizing the operator gestures.

7.3.2.4 Results

The final outcome of the Freestyle use-case was a scenario where a human operator and a mobile robot successfully worked together to assemble a pentomino puzzle, using only gestures to communicate. During evaluation, most target metrics were reached (or surpassed) that provided us a second place in the evaluation (out of five).

The challenges provided by this use-case were centred around the development of pentomino detection, puzzle assembly planning and HRI system. The integration of these technologies into a working system was simplified by our software architecture due to high reusability of already developed components for the Benchmark use-case.

The Freestyle use-case is significantly different from the showcase and field test use-cases, but demonstrates common functionalities such as perception, manipulation, planning, multimodal interaction and cooperation. In fact, detecting piles of blanks, picking, transporting and placing them in the magazine (summary of our showcase/field tests experiment) is a task that has the same structure as detecting, picking, transporting and placing pentominoes.

7.3.3 Showcase

The Showcase use-case was performed in an environment that, although not located in an industrial plant, includes the most important elements of the real environment where the task was supposed be executed as a final product. It uses real pallets with real piles of blanks and a prototype of the blank feeding mechanism of a packaging machine that includes all relevant features for the task to be performed. In the industrial setting, the distance between the blank feeding mechanism and the pallets may be larger, there may be several packaging machines that need to be fed and several types of blanks that must be manipulated.

Our team placed a strong effort into reusing the standard KUKA platform as far as possible without the need of major alterations. However, it was clear from the beginning that a specific gripper was going to be needed. The solution for the showcase resulted from a trial and error approach emerging from a close cooperation between both partners (IMA and UAVR) that, through a few production iterations, converged to the final version of the gripper used for Showcase evaluation. The redesigned gripper includes only passive parts attached to the original gripper mounted in the KUKA platform.

7.3.3.1 Pallet and Blank Perception

Precise and robust pose estimation of blank is a necessary requirement to reliably pick a blank pile from tightly-aligned piles on a pallet (Fig. 7.12). In general all blanks in a pallet have the same shape and size. However, the printed patterns may vary. Therefore, the detection and pose estimations of blanks do not rely on conventional approaches such as color segmentation or local features matching such as SIFT [16]. To detect the blanks we propose a system composed of two modules: detection of a pallet and blank piles and pose estimation of a blank pile.

Since the exact location and configuration of the pallets are not known in advance, the mobile robot moves to a workbench in order to detect them. By using the depth information from the stereo camera installed on the pan-and-tilt unit, the detection of the pallets consists of three sequential steps: first, a voxel grid filter, based on the pallet's height is used to remove irrelevant points; second, the filtered points are projected on an horizontal plane and then a set of clusters is extracted to define candidate regions; and finally, the obtained clusters are compared against 2D template of a pallet projection to find its position and orientation. This same procedure is used to estimate of the piles positions by filtering the point cloud on the average height of the blank piles on the pallet.

Fig. 7.12 Two pallets and blank piles and their detection results. The red lines in center and right images indicate the pose of pallets, while the green circles indicate the positions of blank piles

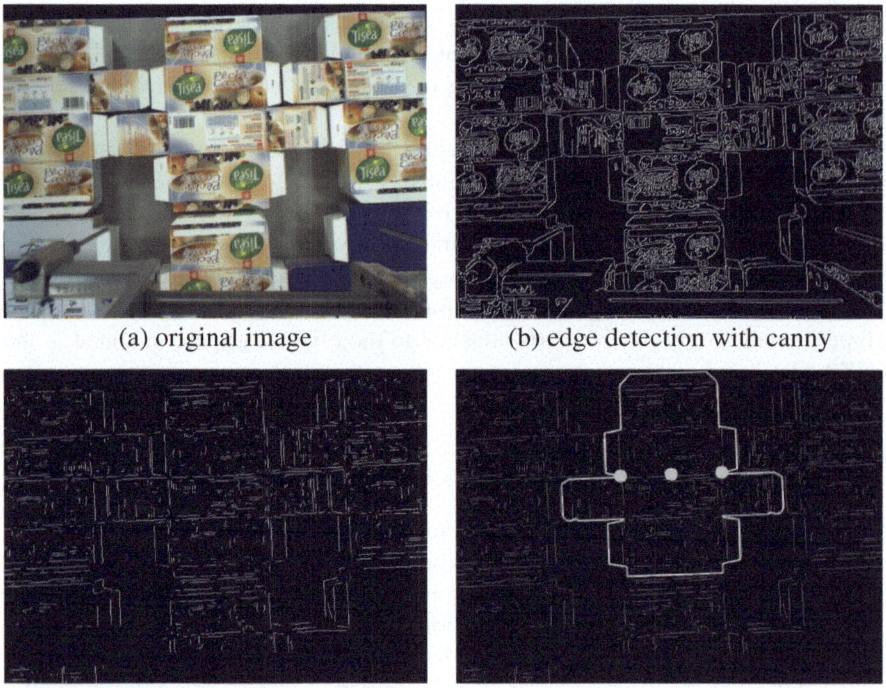

(a) original image (b) edge detection with canny

(c) outline detection with a pin filter (d) blank template matching

Fig. 7.13 A sequential result of the pose estimation of a blank pile

Based on the initial estimated position of pallets and blank piles, the camera attached to the arm gripper is positioned to look vertically at the blank piles and capture an image (see Fig. 7.13a). Then, an edge detector (i.e. canny) is used to identify all edges in the images (see Fig. 7.13b). The resulting edges contain not only the edges from the blank outlines but also lines from inside the blank due to the printed patterns in the blanks. To reduce the edges that derive from the printing patterns, a pin filter is applied (see Fig. 7.13c). The kernel of the filter is defined as

$$K_{DP} = [x_i] = \begin{cases} \frac{p+h}{n}, & \text{if } x_i \text{ is in the center} \\ \frac{h}{n}, & \text{otherwise} \end{cases}$$

$$p > h \times \|[x]\|,$$

$$n = \sum x_i,$$

where p and h are the values of pin and head, respectively. The p is larger than the sum of the heads and n is the normalization factor. In the implementation the pin filter has a kernel of size 5×5 with $p = 50$ and $h = 1$. The reasoning is that the edges of outlines are straight lines and their points have a few number of neighbors while the printed patterns have many neighbors. Finally, to find the pose of the blank, the final filtered image is matched against a scaled template of the blank border, starting by checking the estimated position and then following a spiral path for the next test positions of the template until the threshold for blank detection are surpassed (see Fig. 7.13d).

7.3.3.2 Motion Planning and Manipulation

The blank pile to be manipulated is not a single rigid object, as shown in Fig. 7.14. It is a stack of card boards that not only bends under its own weight but blanks can easily break free from the pile due to the fact that blanks have low friction between each other. Any manipulation procedure of the a blank pile has to be aware of this fact. After picking, the pile has to be transported to the feeding magazine of the packaging machine. In the end, the blank pile has to be placed in the feeding magazine avoiding physical collision with the blanks that are already in place and with the four guiding rods that prevent blanks from slipping out of the inclined magazine surface.

By using a single arm with a gripper, the "picking" of a pile required, in the Showcase, three manipulation steps. Depending on the arrangement of the blank pile on the pallet, the robot needs to drag a blank pile over the edge of the pallet so there is enough free space for grasping without causing a breakdown or losing any blank. Then it has to grasp the pile from the edge and finally it has to recondition the pile for a firm grip. After grasping the blank pile, it is necessary to move the arm to the transport position, which is inside the mobile platform footprint, while having the arm in a pose that minimizes the necessary motions to place the blank pile in the feeding magazine. All manipulations have to be done while maintaining the blank pile in a pose that prevents it from breaking lose. This is a constraint that has to be accounted for when planning the motions of the robotic arm. To achieve the best motion trajectories that respected our constraint, several different motion planner were tested and benchmarked [17].

Due to the disposition of the blank piles on the pallet, certain piles are blocked by others and can not be manipulated. Therefore, the order in which the blank piles

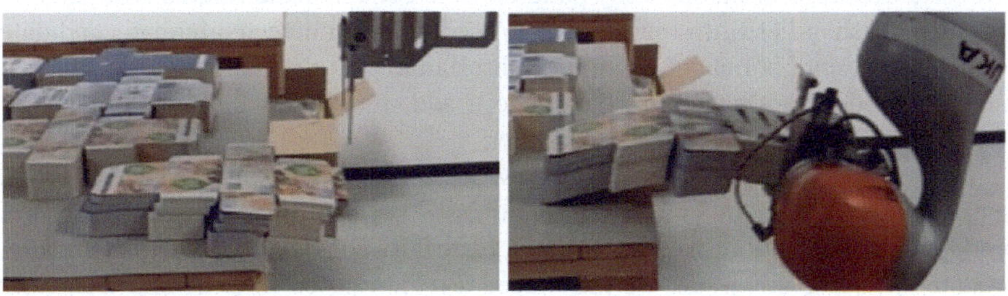

Fig. 7.14 Snapshots of bent blank piles

are manipulated has to be planned. The planner considers the available blank piles on the available pallets and gives preference to the pallet with the lowest number of piles.

7.3.3.3 Safe Human-Robot Collaboration

Since industrial environments are noisy, verbal communication with human can be challenging and impractical with robots. Thus, gestures were considered as a practical alternative method of communication. The communication with the robot is therefore through gestures using the HRI system developed for the Freestyle use-case and reused for Showcase with changes only to interaction graph.

In the Showcase use-case, the human operator is no longer assumed to be confined to a single location. Now, the mobile platform and the operator can share the same physical workspace. In fact, it is expected that a human operator may physically approximate the robot in order to provide commands through gestures. This possible event raises safety concerns that result in an active and continuous monitorization of humans. The physical proximity of a human must result in a halt of moving parts of the mobile platform and, at the same time, it should position the relevant sensors in order to interpret possible order from the operator [18].

7.3.3.4 Human Detection and Tracking

The vision system integrated in the mobile platform (two stereo camera installed on the gripper and pan-and-tilt unit) have a limited field of view and shared access conflicts, i.e. if a camera is used for blank detection it can not be used for human detection. Therefore, to detect and track humans constantly, two laser range finders (LRF) were mounted on both sides of the mobile platform to cover all directions with a reduced area of shadow regions (i.e. areas that are no visible by the sensors). The multiple single-plane laser scans are merged to generate a virtual laser scan with a 360° field of view.

The detection of a human is performed with a machine-learning-trained classifier that receives the virtual laser scan as input [18]. The positive and negative samples of laser scan readings of human body patterns were collected and trained with random forests beforehand [19]. In the tracking procedure, each detected human candidate in the virtual scan is matched, within a threshold, with the list of already tracked human. If it does not match, the candidate is added to the tracking list. Finally, all tracks are updated based on a Kalman filter.

The working environment of the Showcase adds an additional challenge to the human tracking system. The natural features of the environment, such as from a workbench, the robotic arm or blank piles, create false positives in the human tracking module. To solve this issue, we introduced a region-based filtering and reasoning mechanism [18]. Most false positives arise from areas where the humans can not be physically, so those areas are marked as "black" rectangles and detection inside that areas are removed from the human detection. Additionally, a "green" rectangle defines the regions where human detection candidates are considered.

7.3.3.5 Safety Management

Safety is handled using the output of the human tracking module combined with a set of working regions, which are defined by the region-based filtering and reasoning mechanism. Two rectangular regions are defined around the mobile platform during navigation. An inner rectangle (a "red" rectangle) is close to the robot and an outer rectangle (a "yellow" rectangle) that covers a wider area. During arm manipulation, a circular region (a "red" circle) is defined around the arm, which radius is determined by the range of the arm. For the circular region, only the areas of the circle that do not intersect with the inner rectangle are considered.

Based on the fixed regions defined for human tracking and the dynamic regions around the mobile platform, the safety handle proceeds as follows. Any human candidate in the "green" region alone, is considered safe, i.e., the robot operation does not put any human in a "green" region in danger of physical harm. Therefore, no safety action is taken. If a human enters a "yellow" region, the navigation is suspended but the arm keeps doing its current operation. If a human enters a "red" region, then both the mobile platform and the arm are suspended until the human leaves the region in question. Both "yellow" and "red" regions also trigger HRI systems. HRI will take control of the pan-and-tilt unit to redirect the stereo camera to the human operator and awaits for a command every time "yellow" or "red" regions have humans. If the operator leaves the region of interest then the pan-and-tilt is released.

7.3.3.6 Scenarios

The Showcase use-case execution consisted of three different scenarios with increasing complexity. Here are their description:

Scenario 1 The goal of this challenge was to transport 2 blank piles to the feeding magazine. The robot had to identify the feeding magazine, the pallet and the piles.

Scenario 2 A second pallet is introduced to the task. The robot has to show that it selects the pallet with the lowest number of blanks first. Additionally, it has to switch pallets when needed and navigate accordingly.

Scenario 3 In this challenge the human operator was introduced. Now, the robot had to demonstrate safety features needed when humans share the robot environment complemented with human-robot interactions.

For the purpose of evaluation, several metrics had to be defined. Those include metrics for: perception, where the blanks the pallets and the feeding magazine had to be correctly detected; manipulation, where the manipulation of the blank piles had to be done correctly with safety concerns; and interaction, where the robot has to successfully track any possible human operator and be able to communicate.

7.3.3.7 Results

The final outcome of the Showcase use-case was a scenario where a human operator and a mobile robot successfully work together to pick and place piles of blanks. All target metrics were reached with success which provided us the first place in the evaluation (out of five). The reusability of previously developed technologies is an important aspect for the success of our solution. This aspect facilitates integration of different technologies and facilitates the rapid development of new and also flexible solutions.

The Showcase evaluation was performed in an environment that, although not located in an industrial plant, includes the most important elements of the real environment where the task will be executed as a final product. It uses real pallets with real piles of blanks and a prototype of the blank feeding mechanism of a packaging machine that includes all relevant features for the task to be performed. In the industrial setting, the distance between the blank feeding mechanism and the pallets may be larger, there may be several packaging machines that need to be fed and several types of blanks that must be manipulated.

7.3.4 Field Tests

The Field Tests use-case is an evolution of the Showcase that takes us closer to Autonomous Blank Feeding for Packaging Machines in a industrial environment (Fig. 7.15). TIMAIRIS field tests use-case was composed of a shop floor version of the complete blank feeding problem. It included: two packaging machines using different blanks and with different feeder mechanisms at their blank magazines; the mechanism that feeds the blanks into the packaging machine; and three pallets filled in with 15 cm height blank piles used in production. The task consisted in recognize the blank piles on the three pallets, devise a plan for picking them considering the packaging machines blank consumption rates and current blank level at each magazine, and finally transport and place them in the blank magazines.

7.3.4.1 General Architecture

The Field Tests use-case introduced the online monitoring of the blank feeding level, which is provided by external components to the mobile platform. Also, there is an external dashboard that provides relevant information to the operator. The introduction of these external components expanded our architecture, centered in the mobile platform, to a factory-wise architecture. The architecture is depicted in Fig. 7.16.

For the Field Tests the robot platform was enhanced to include a new gripper device that includes a parallel 2-jaw gripper from Zimmer with a higher gripping force that enables manipulation of 15 cm high piles. The new gripper mechanical fingers allow picking the blank pile from the pallet in the vertical direction. Another

Fig. 7.15 Field Tests environment layout. It includes three pallets and 2 machines with blank feeders

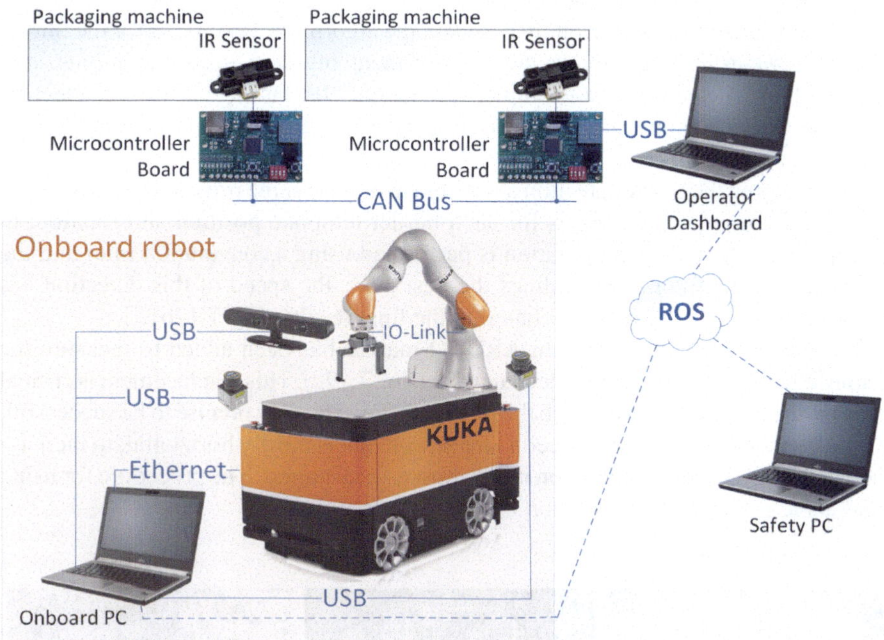

Fig. 7.16 TIMAIRIS architecture used in the field tests use-case

enhancement was the inclusion in the robot platform of a pile buffer that can be used to store a pile, while the other is being transported by the gripper, and a rotation buffer, used to re-grasp the piles in order to be able to place them at the feeders.

The LRFs used for human tracking in the Showcase have also been used in the Field Tests, as well as the Asus Xtion, that has been used mainly for human-robot interaction. An onboard PC runs the solver and is connected to ROS communication network.

IR distance sensors have been added to the feeders of both packaging machines to measure the blank level. These sensors are managed by a microcontroller board and connect to the operator dashboard (a laptop PC running ROS) using a CAN bus. This architecture is prepared for the expansion of the number of packaging machines. The safety PC, that runs the safety approach and human tracking procedures, is an external laptop, that communicates with others through ROS.

7.3.4.2 Perception

Field Tests perception was strongly based on the algorithms developed for the Showcase. The main changes are in the improvement of some implementations, that allowed blank detection, for example, to become 10x faster, and in the selection of the observation points for pallet and magazine detection that had to be adapted to the new environment (Fig. 7.17a).

By changing the blank detection algorithm, using the same principles, from counting the number matching pixels for each border template position and rotation, to an approach where this computation is performed using a convolution filter and the maximum of the filter result defines the best pose, the speed of this detection has been increased by 10x, without changing the final results (Fig. 7.17b).

Near each blank magazine, an ARUCO marker has been added to measure the displacement from the robot to the magazine (Fig. 7.17c). This displacement is critical because the manipulation to place the blank pile must be very precise to be successful. ARUCO markers have been placed in a position that is mostly horizontal, to increase the sensitivity of their localization in the most critical angles. The procedure for using markers to perform this correction has been generalized.

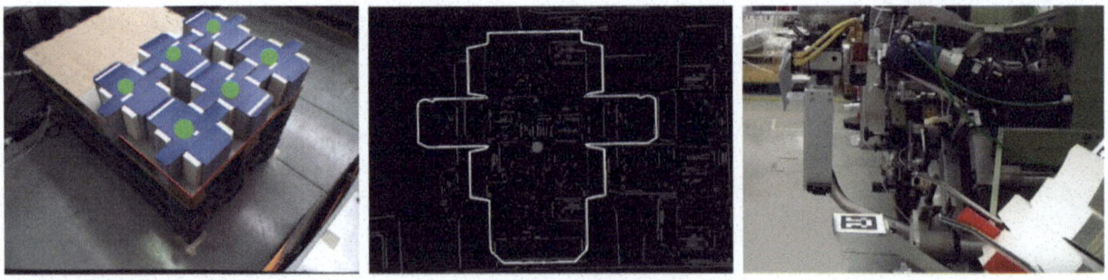

(a) pallet and pile detection (b) blank detection (c) marker detection

Fig. 7.17 Examples of perception in the Field Tests use-case

7.3.4.3 Planning

In the Field Tests use-case, the robot had to feed blanks at 2 packaging machines. In addition, the blanks shapes for these 2 machines were different and the blank magazines had different capacities and restrictions for blank loading. In order to get a general solution which can be adapted to the number of packaging machines and to their characteristics, the sequence of operations was derived using a PDDL 2.1 planner (named popf) [20] for high-level actions and a low-level planner for choosing the picking blank pile within the pallets. The high-level actions that were considered were the following:

- `move ?from ?to`
- `pick ?pile ?waypoint ?gripper`
- `pick-to-buffer ?pile ?waypoint ?gripper`
- `place-big ?pile ?waypoint ?gripper`
- `place-small ?pile ?waypoint ?gripper`
- `place-from-buffer ?pile ?waypoint ?gripper`
- `human-disarm ?pile`

where `big` and `small` designate the 2 types of blanks that were used in the Field Tests. The parameters of the planning problem are:

- number of packaging machines
- type of blanks for each packaging machine
- initial position of robot and available blank piles
- initial blank level at each blank magazine
- rate of consumption of blanks for each packaging machine
- highest and lowest blank levels that allow inserting a new pile in each magazine
- estimated duration of each high level operation.

The problem of planning is rather complex as the environment is not stationary. The packaging machines are always working, decreasing the blank level at their magazines, while the robot is in operation. The plan should make sure that all machines are fed with blanks, considering the restrictions of the blank magazines, without any of them getting at the lowest level, to ensure continuous operation.

The domain of this problem was modelled using PDDL 2.1 with durative-actions. To model the non-stationary part of the environment, PDDL functions for the magazine level of each magazine were used. The value of these functions is updated for every action, according to the action duration. Some step actions had to be inserted to model periods where the robot must wait without performing any of the high-level actions. These step actions keep the magazine levels updated.

The planning problem is implemented by a ROS node that implements the service *get_plan*. This service reads the magazine levels, detected by the IR sensors at each magazine, considers the blank piles of each type present at the pallets and generates the PDDL problem description. Planning is performed using the popf planner from the ROSPlan framework [21]. The service continues to parse the plan generated by

Fig. 7.18 Snapshot of the the robot following the first step of the execution plan

popf, publishes the plan and makes it part of the service response. Figure 7.18 presents the robot executing the first action of the plan that is shown in the top of the figure. A graphical user interface has also been developed that allows continuous monitoring of the blank levels, a visualization of the current plan, and an offline planning mode where the plan parameters can be edited interactively, instead of being read from the environment, and that was used for testing the planner (Fig. 7.19).

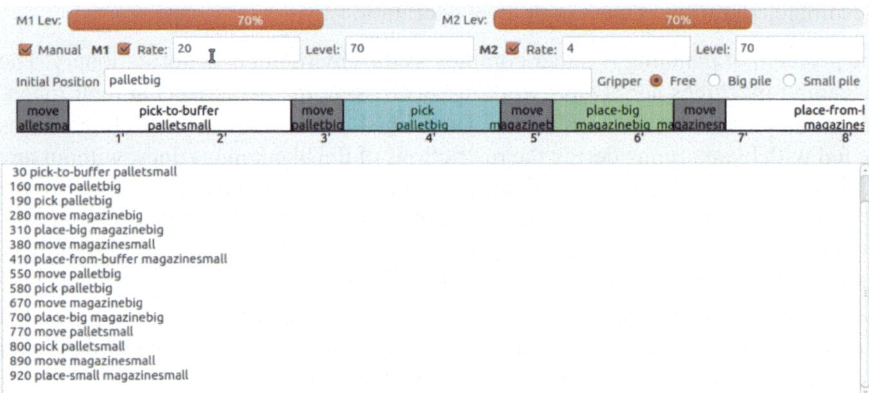

Fig. 7.19 Execution dashboard. Its show the state of the feeding magazine level, the current plan and an execution log. It also allows to set several execution parameters

7.3.4.4 Manipulation and Safety/Interaction

Field Tests manipulation had to be re-engineered from the approach used in the Showcase due to the different gripper. Additional changes to manipulation are the actions performed by the robot arm to put the pile in the rotation buffer and pick it again, as well as the placing of the pile in the transport buffer. Nonetheless, most manipulation actions are similar to the ones performed in the Showcase, thus reducing the development time. Example of manipulation action performed in the Field Tests use-case are shown in Fig. 7.20.

The safety/interaction part of the Field Tests use-case is the same from the Showcase. The difference is that now multiple humans can be present in the environment. Now the safety management assumes that the human operator is the one closest to the pan-and-tilt unit. An example of interaction and human tracking is shown in Fig. 7.21.

7.3.4.5 Scenarios

TIMAIRIS Field Tests evaluation was structured around 3 different scenarios of increasing complexity. These scenarios play a similar role as the challenges used for the freestyle and showcase use-cases.

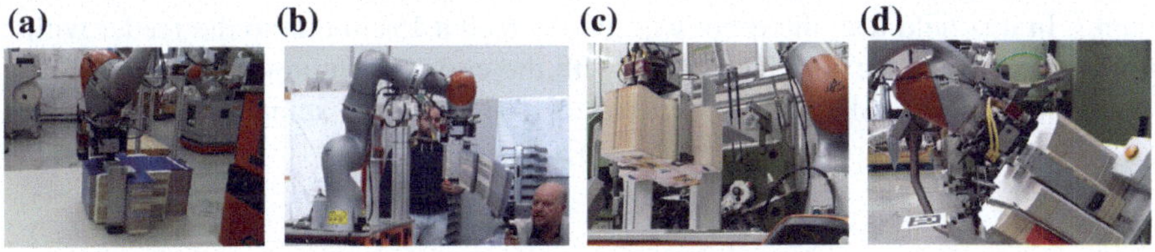

Fig. 7.20 Example of manipulation actions used in the Field Tests use-case. **a** Picking with compliance. **b** Picking of non-rigid objects. **c** Using the pile buffer. **d** Smooth placing in the feeding magazine

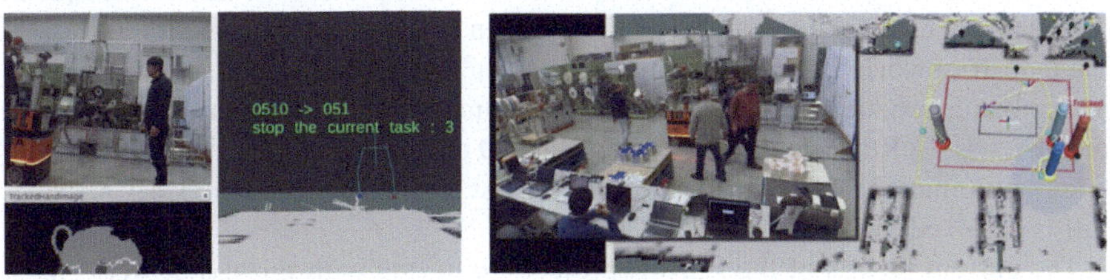

(a) gesture recognition (b) human tracking and safety regions

Fig. 7.21 Example of human-robot interaction and human tracking

7.3.4.6 Results

The Field Tests have demonstrated the feasibility of practical use of an intelligent mobile robot in this scenario, by exploiting the following technologies: skill-based anytime agent architecture, in which the agent can start execution at any task state; knowledge-based perception, that enables the robot to both quickly detect the pose of the pallets and blank magazines using spatial knowledge, and to recognize the pose of densely and closely lying blank piles; a set of skills and a sophisticated gripper which enable the robot to grasp and adjust an untied pile of non rigid blanks, move the arm into a safe navigation posture and to smoothly feed the pile to two different complex shaped blank magazines; planning and re-planning, which produces a sequence of actions for manipulation; rich multimodal interaction which enables the robot to interact with human collaborators in a noisy environment by encoding a set of commands with a fixed number of symbols.

The Field Tests workspace mimics the conditions that will be found in a real scenario. Efficient navigation on reduced spaces, two complete packaging machines with two different blank feeding devices, including the physical limitations imposed by the mechanical cluttering, non deterministic location of the blank pallets, changing illumination conditions and the presence and interaction with human co-workers. The excellent results obtained during the field tests phase, which did not focus on speed efficiency, can in fact demonstrate in a clear and provable way that the final prototype will be able to work under realistic conditions. Note also that the typical feeding rate of a packaging machine is around 1 cm/min for large blanks and 0.8 cm/s for small blanks. In this field test, the robot was able to feed a 15 cm pile to the feeder within a maximum time of 2 min. Therefore, with the current solution, not optimized for speed, the robot would already be able to feed four packaging machines, which confirms the results provided by the planner.

7.4 Conclusion

IMA's position of world leadership in the production of packaging machines requires a constant focus on staying ahead of the curve in terms of technological innovation. The industrial world is looking more and more at automation as a resource to manufacture better products in a cost-efficient way, and this area is of utmost importance for IMA, too. EuRoC is not the only IMA's R&D project on robotics and automation; in fact, automation is one of the two/three most important areas the IMA research staff is involved in.

The current results of the EuRoC project are in line with those that were stated at the project submission. The convenience of robotised features reflects into customers', workers', and sustainability benefits. From the customers' perspective, robotised systems would make it possible to avoid the use of a human operator dedicated to blank loading, resulting into cost cutting and efficiency gaining. This also would give the operator more time for other operations such as quality control, or looking after more machines. From the workers' point of view, the adoption of robots to automate repetitive procedures is more than welcome. Such robotised systems alleviate the stress derived from repetitive labour and contribute to build real factories for humans.

As for sustainability, adopting robotised systems also affects the sustainability of the entire packaging process. In fact, fewer mistakes in feeding wrong blanks, or feeding them in the wrong way, results into less waste, fewer machine stoppages, and a reduced energy consumption—a mandatory requirement for a green factory.

TIMAIRIS is confident that the overall results of the project—and of IMA's commitment to robotics and automation—will manifest themselves in a stronger leadership position, harder and harder to overcome, and in a wider presence on the international markets. Moreover, the EuRoC project has given IMA the opportunity to develop links with research centers with which to find innovative solutions to automation issues. This type of knowledge will be deepened and used to find innovative solutions to many other problems. IMA and UAVR signed a contract, funded by IMA, to continue their collaboration during 2018/2019.

It is difficult to imagine the future of industry and manufacturing without a massive role being played by robotics and automation. IMA is actively working on this topic, both in the framework of the EuRoC project and in other R&D projects. The results of this wide range of activities will constitute the backbone of a new way of designing machines that employ robots and automation in general, the future IMA's products will have to be designed around the notion that an increasing number of repetitive and low-skilled actions will be carried out by machines.

The EuRoC project will bring IMA to develop new products that support and complement its automatic machines, as well as new services associated to cooperative robotised platforms. Also to be stressed is that the use of this kind of robotised blank feeding is intended to be included as an option in future packaging machine products as soon as its robustness reaches the required Technology Readiness Level.

Overall, the EuRoC project had a strong impact both on UAVR and IMA. EuRoC provided the framework to start the collaboration between these 2 institutions and this collaboration is active and effective after EuRoC is finished. The complementarity of UAVR and IMA is well represented in Fig. 7.22.

Some of the strong points in the EuRoC project, from the TIMAIRIS perspective, are its mandatory cooperation between academia and industry focused on specific challenging applications; the structure of EuRoC competition (Stages I, II and III), their respective tasks, the rank based filtering of teams for entering new stages and the provided funding worked very well to foster a productive environment. The interaction between TIMAIRIS and Challenge 2 host, and the tools, including ROS wrapper, eased-up the developments and the exchange of ideas and feedback from host was always very important; the meetings and workshops that joined all EuRoC teams, core partners and evaluators was very enriching.

Some points that, from TIMAIRIS perspective, could be improved are: some of the evaluation models should be thought more carefully, as for example allowing some metrics to dominate the evaluation in Stage II; midterm evaluation workshop format was not clear (great improvement was performed for Showcase and Field Tests

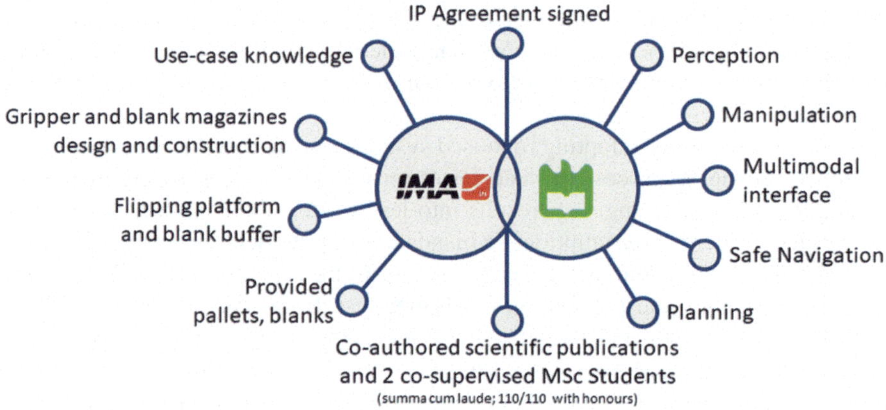

Fig. 7.22 Complementarity between UAVR and IMA

workshops); some deadline dates and penalties for missing them were not clear or conveyed to the teams with a short notice.

TIMAIRIS would especially like to acknowledge the specification of the succession of Benchmarking and Freestyle (Stage IIa); Showcase (Stage IIb) and Field Tests (Stage III) and their competitive/selective nature that provided a very good ramp up for the development towards a successful solution for the industrial problem to be solved. The Benchmarking was a very important tool to get to know the EuRoC Challenge 2 robot platform capabilities, and also its limitations; increase the mutual knowledge and trust between UAVR and IMA and also between these 2 institutions and Challenge 2 host (DLR). TIMAIRIS Freestyle specification was provided by TIMAIRIS in its Stage II proposal. This specification was not directly related to IMA's industrial problem, but it provided for the development of many technologies that were used in the next stages. The freestyle objectives were placed at a very high level, and were very well demonstrated during the evaluation. The Showcase provided for a first implementation of autonomous blank feeding in a lab setup. Finally, the Field Tests allowed the development of solutions in an industrial environment, very close to the final product.

We would like to thank the EuRoC Core partners, and also Challenge 2 host, for their effort in the EuRoC specification and management and for providing the framework for a very productive TIMAIRIS research and development work.

Acknowledgements This work was supported by the EuRoC Project under Grant no. 608849 and by Portuguese National Funds through the FCT—Foundation for Science and Technology, in the context of the project UID/CEC/00127/2019.

References

1. Hermann, M., Pentek, T., Otto, B.: Design principles for industrie 4.0 scenarios. In: 2016 49th Hawaii International Conference on System Sciences (HICSS), pp. 3928–3937. IEEE (2016)
2. Bischoff, R., Guhl, T.: The strategic research agenda for robotics in Rurope [industrial activities]. IEEE Robot. Autom. Mag. **17**(1), 15–16 (2010)
3. Siciliano, B., Caccavale, F., Zwicker, E., Achtelik, M., Mansard, N., Borst, C., Achtelik, M., Jepsen, N.O., Awad, R., Bischoff, R.: Euroc—the challenge initiative for European robotics. In: ISR/Robotik 2014; 41st International Symposium on Robotics, pp. 1–7. VDE (2014)
4. Pedrocchi, N., Vicentini, F., Matteo, M., Tosatti, L.M.: Safe human-robot cooperation in an industrial environment. Int. J. Adv. Robot. Syst. **10**(1), 27 (2013)
5. Quigley, M., Conley, K., Gerkey, B., Faust, J., Foote, T., Leibs, J., Wheeler, R., Ng, A.Y.: ROS: an open-source robot operating system. In: ICRA Workshop on Open Source Software, vol. 3, p. 5. Kobe, Japan (2009)
6. Pedrosa, E., Lau, N., Pereira, A., Cunha, B.: A skill-based architecture for pick and place manipulation tasks. In: Pereira, F., Machado, P., Costa, E., Cardoso, A. (eds.) Progress in Artificial Intelligence, pp. 457–468, Coimbra, Portugal, September 2015. Springer International Publishing
7. Amaral, F., Pedrosa, E., Lim, G.H., Shafii, N., Pereira, A., Azevedo, J.L., Cunha, B., Reis, L.P., Badini, S., Lau, N.: Skill-based anytime agent architecture for logistics and manipulation tasks: EuRoC Challenge 2, Stage II—Realistic Labs: Benchmarking. In: 2017 IEEE International Conference on Autonomous Robot Systems and Competitions, ICARSC 2017, pp. 198–203 (2017)
8. Rusu, R.B., Marton, Z.C., Blodow, N., Dolha, M., Beetz, M.: Towards 3D point cloud based object maps for household environments. Robot. Auton. Syst. **56**(11), 927–941 (2008). Semantic Knowledge in Robotics
9. Rusu, R.B.: Semantic 3D object maps for everyday manipulation in human living environments. Ph.D. Thesis, Computer Science department, Technische Universitaet Muenchen, Germany, October 2009
10. Hamidreza Kasaei, S., Seabra Lopes, L., Tomé, A.M., Oliveira, M.: An orthographic descriptor for 3D object learning and recognition. In: 2016 IEEE/RSJ International Conference on Intelligent Robots and Systems (IROS), pp. 4158–4163 (2016)
11. Pedrosa, E., Pereira, A., Lau, N.: A scan matching approach to SLAM with a dynamic likelihood field. In: 2016 International Conference on Autonomous Robot Systems and Competitions (ICARSC), pp. 35–40, Portugal, Bragança. IEEE (2016)
12. Lim, G.H., Pedrosa, E., Amaral, F., Lau, N., Pereira, A., Dias, P., Azevedo, J.L., Cunha, B., Reis, L.P.: Rich and robust human-robot interaction on gesture recognition for assembly tasks. In: 2017 IEEE International Conference on Autonomous Robot Systems and Competitions, ICARSC 2017, pp. 159–164 (2017)
13. Gleeson, B., MacLean, K., Haddadi, A., Croft, E., Alcazar, J.: Gestures for industry: intuitive human-robot communication from human observation. In: Proceedings of the 8th ACM/IEEE International Conference on Human-Robot Interaction, pp. 349–356. IEEE Press (2013)
14. Lim, G.H., Suh, I.H.: Robust robot knowledge instantiation for intelligent service robots. Intell. Serv. Robot. **3**(2), 115–123 (2010)
15. Norvig, P.: Natural language corpus data. In: Beautiful Data, pp. 219–242 (2009)
16. Lowe, D.G.: Object recognition from local scale-invariant features. In: The Proceedings of the Seventh IEEE International Conference on Computer Vision, vol. 2, pp. 1150–1157. IEEE (1999)
17. Tudico, A., Lau, N., Pedrosa, E., Amaral, F., Mazzotti, C., Carricato, M.: Improving and benchmarking motion planning for a mobile manipulator operating in unstructured environments. In: Portuguese Conference on Artificial Intelligence, pp. 498–509. Springer (2017)
18. Lim, G.H., Pedrosa, E., Amaral, F., Dias, R., Pereira, A., Lau, N., Azevedo, J., Cunha, B., Reis, L.P.: Human-robot collaboration and safety management for logistics and manipulation tasks.

In: Ollero, A., Sanfeliu, A., Montano, L., Lau, N., Cardeira, C. (eds.) Advances in Intelligent Systems and Computing, vol. 694, pp. 15–27. Portugal. Springer International Publishing, Porto (2017)
19. Breiman, L.: Random forests. Mach. Learn. **45**(1), 5–32 (2001)
20. Coles, A., Coles, A., Fox, M., Long, D.: Forward-chaining partial-order planning. In: Proceedings of the Twentieth International Conference on International Conference on Automated Planning and Scheduling, ICAPS'10, pp. 42–49. AAAI Press (2010)
21. Cashmore, M., Fox, M., Long, D., Magazzeni, D., Ridder, B., Carreraa, A., Palomeras, N., Hurtós, N., Carrerasa, M.: ROSPlan: planning in the robot operating system. In: Proceedings of the Twenty-Fifth International Conference on International Conference on Automated Planning and Scheduling, ICAPS'15, pp. 333–341. AAAI Press (2015)

Part IV
Achievements in Challenge 3: Plant Servicing and Inspection

Introduction

Zachary Taylor

Micro Aerial Vehicles (MAVs) are becoming an increasingly valuable tool for performing inspection, maintenance and logistics tasks in industrial environments. Tasks that once required people to work in difficult or dangerous locations and significant investment in equipment can now be performed quickly and efficiently by these systems. For example, the inspection of tall infrastructure can be performed without the need to erect scaffolding or requiring individuals to work at heights. This is thanks to the freedom of movement the MAV systems provide in addition to their flexibility and independence from external infrastructure. However, at present, these systems require skilled operates to use efficiently, and tasks such as hovering close to infrastructure in GPS-denied locations remain challenging. They are also typically only equipped with a single camera and are limited to the task of recording or streaming video or photos to the user, with the processing and understanding of the provided footage remaining a completely manual task. This processing can take significant time and again requires that the user is an expert in the material present to them. These limitations have slowed the rate of adoption of this technology, preventing its use in many applications where it has the potential to increase safety while simultaneously reducing cost.

To address these issues, Challenge I3 had two key objectives: first, to automate the low-level control of MAV systems to allow untrained operators to make use of them through simple high-level interfaces and decisions, and second, to equip the MAVs with a range of sensors that would allow them to go beyond simple waypoint following and allow the system to perceive and interact with its environment, performing such actions as 3D scene reconstruction, obstacle avoidance and area of interest detection.

The teams that participated in this Challenge covered a wide range of use case scenarios. This chapter will outline the application the two finalist teams chose to take on, their approach and details of their system and results. The first finalist was

TUM Flyers, who looked at the task of automating bridge inspection, with the goal of mapping a designated area of a bridge and providing high-quality images of its surface. In addition to this, automated crack detection was performed on the gathered images.

The second finalist, team GRVC-CATEC, examined the problem of tool delivery and lost tool localization in an Airbus factory producing cowlings for jet engines. Both teams demonstrated their systems in collaboration with an end user and performed live demonstrations at the end user site. The situations required the deployment of an MAV that had the capacity to localize and automatically move through a location with unreliable or no GPS and had the goal of automating what is currently a highly manual task.

Chapter 8
TUM Flyers: Vision—Based MAV Navigation for Systematic Inspection of Structures

Vladyslav Usenko, Lukas von Stumberg, Jörg Stückler and Daniel Cremers

Abstract A large amount of buildings and structures in Europe's industry and infrastructure require regular inspection and maintenance. Inspection nowadays often is a costly and time-consuming process. Inspection teams need to climb through the structures, or special heavy equipment such as access lifts need to be employed. Current approaches using micro aerial vehicles (MAVs) rely on GPS for autonomous navigation and require highly skilled and well trained pilots to maneuver the MAV close to structures. It is often difficult for the team to perform a full and systematic inspection, which may cause deficiencies to be overlooked. Finally, the inspection team typically has no accurate position information to exactly reference deficiencies, such that comparisons across multiple inspections are difficult to achieve. The aim of our project is to develop novel vision-based localization, 3D reconstruction, and navigation technologies for increasing the level of autonomy of MAV inspection systems and the systematic quality of inspections. The inspection with MAVs of a large structure such as a bridge is a highly challenging real use-case for which we have developed a feasibility demonstration in this project. Our scenario combines a variety of open problems for the vision-based localization, mapping and autonomous navigation for MAVs in large-scale outdoor scenarios, which are shared with many other tasks like the inspection of buildings, industrial plants, wind power plants, or oil platforms. The technologies pursued in this project target at further automating

V. Usenko (✉) · L. von Stumberg · D. Cremers
Technical University of Munich, Boltzmannstr. 3, 85748 Garching, Germany
e-mail: vlad.usenko@tum.de

L. von Stumberg
e-mail: stumberg@in.tum.de

D. Cremers
e-mail: cremers@tum.de

J. Stückler
Max Planck Institute for Intelligent Systems, Tübingen, Germany
e-mail: joerg.stueckler@tuebingen.mpg.de

F. Caccavale et al. (eds.), *Bringing Innovative Robotic Technologies from Research Labs to Industrial End-users*, Springer Tracts in Advanced Robotics 136,
https://doi.org/10.1007/978-3-030-34507-5_8

and simplifying the use of MAVs in GPS-restricted or GPS-denied environments. These technologies have the potential to increase cost effectiveness and the quality of inspection and maintenance operations.

Keywords MAV · UAV · Inspection · Localization · Mapping · SLAM

8.1 Introduction

A large amount of buildings and structures in Europe's industry and infrastructure require regular inspection and maintenance. Inspection nowadays often is a costly and time-consuming process. Inspection teams need to climb through the structures, or special heavy equipment such as access lifts need to be employed. Current approaches using micro aerial vehicles (MAVs) rely on GPS for autonomous navigation and require highly skilled and well trained pilots to maneuver the MAV close to structures. It is often difficult for the team to perform a full and systematic inspection, which may cause deficiencies to be overlooked. Finally, the inspection team typically has no accurate position information to exactly reference deficiencies, such that comparisons across multiple inspections are difficult to achieve. The aim of this project is to develop novel vision-based localization, 3D reconstruction, and navigation technologies for increasing the level of autonomy of MAV inspection systems and the systematic quality of inspections. The inspection with MAVs of a large structure such as a bridge is a highly challenging real use-case for which we have developed a feasibility demonstration in this project. Our scenario combines a variety of open problems for the vision-based localization, mapping and autonomous navigation for MAVs in large-scale outdoor scenarios, which are shared with many other tasks like the inspection of buildings, industrial plants, wind power plants, or oil platforms. The technologies pursued in this project target at further automating and simplifying the use of MAVs in GPS-restricted or GPS-denied environments (Fig. 8.1). These technologies have the potential to increase cost effectiveness and the quality of inspection and maintenance operations.

8.2 Use Case Description and Motivation

Services for inspection and maintenance of structures with MAVs are already offered by several commercial providers such as Ascending Technologies, Microdrones, or Aibotix. These current solutions, however, still require a highly skilled, expert pilot to operate the MAV close to structures. Autonomous flight is typically limited to GPS waypoint navigation, while obstacles are avoided by flying in heights above vegetation and structures. Due to the manual piloting close to structures, an accurate systematic inspection at predefined view poses is difficult to achieve. In this project, we have developed vision-based MAV navigation techniques that assist operators

Fig. 8.1 The bridge selected for field test demonstration (Sitterviadukt St. Gallen Switzerland)

in piloting the MAV. Systematic inspection is enabled by autonomous vision-based waypoint navigation along structures.

Autonomous MAV navigation requires obstacle avoiding control as well as path planning between waypoints. Typical approaches such as [1, 2] are multi-layered: paths are planned on a global scale, which are then followed using local control. In SPARTAN [1], a roadmap is sampled tangentially around obstacles. Search-based planning (i.e. A*) in this roadmap finds paths that avoid obstacles and quickly cut through free-space. The method achieves replanning at 2 Hz for distances of about 40 m. The approach proposed by Achtelik et al. [2] applies a sampling-based rapidly-exploring random tree (RRT) planner to find collision-free global paths. Collision-avoiding smooth polynomial trajectories are fitted to the planned paths and executed with a flight controller. While these methods assume complete knowledge about obstacles through the global map or a full coverage with LIDAR sensors, one novel challenge in our vision-based navigation approach is that we need to cope with a limited field-of-view. This imposes constraints on MAV orientation and velocities during flight, which need to be incorporated into path planning and local control. In our previous works the MAV state is estimated through the fusion of sparse interest-point based visual localization with IMU information. The MAV is locally controlled along trajectories based on the visual state estimate. Besides autonomous waypoint navigation, our system also supports an obstacle-avoiding semi-autonomous flight mode.

We also pursue the visual recognition of local surface changes in structures. Traditional approaches analyse image content for strong edges and their distribution in the image e.g. using Fourier analysis [3]. Machine learning methods have been less frequently applied for this purpose [4]. We also pursue a novel complementary way by detecting changes between inspections. One related approach is change detection in satellite images [5].

Our team consists of the Computer Vision Group (CVG) at the Technical University Munich as a research partner and the end-user Schällibaum AG Ingenieure und Architekten (SLB) which contributes its use-case of bridge inspection. Both partners have significant prior experience with MAVs. CVG has a strong research background in vision-based navigation for MAVs, 3D reconstruction, and semantic interpretation of images. SLB already uses MAV technology for providing services through its geomatics division to generate simple aerial photo- and video-shootings over photogrammetric surveys up to complex inspection images for civil engineering and architectural purposes. So far, SLB pilots its MAVs manually and on GPS waypoints. The technologies pursued in the Showcase potentially enlarge the economic efficiency of SLB 's use-case.

SLB as a provider of engineering and architectural solutions in the construction industry, SLB maintains a lot of contacts to possible customers and knows the requirements of inspection/survey tasks out of the practice. Since 2012, SLB provides MAV survey and inspection as a new service to its customers. In this time, SLB gathered a lot of knowledge about the current state-of-the-art of the MAV technology and its limitations. As a potential end-user and with the experience related to the working environment, SLB has brought the user perspective into the team. The close distance between the home bases of all team members and the demonstration locations facilitated the communication and exchange in the team as well as meetings for joint tests of the developed system and technologies. SLB has shared its knowledge about real use-case requirements with CVG, which was used by CVG for the development of sustainable solutions that could be applicable to SLB' s use-case, other end-users, tasks and situations.

8.3 Activities and Results

8.3.1 Freestyle Stage

For the Freestyle stage, our task was to develop the core technologies required for the structure inspections in GPS-denied environments and demonstrate them by inspecting a structure mocap in the flying arena of the challenge host. These technologies include real-time visual localization and mapping using the MAV's on-board sensors, semi-autonomous and autonomous flight modes. To allow better evaluation of the freestyle it was subdivided into 4 objectives, each of them demonstrating and evaluating different aspects of the system: In the first objective we evaluate accuracy of the visual odometry and 3D reconstruction of the environment. Accuracy of the visual odometry is important for the position-control mode (user can release the controls, and the drone will not drift away) and trajectory following. Accurate generation of 3D maps of the environment is necessary for planning collision-free trajectories. In the second objective we demonstrate an enchased user interface that avoids collisions, even when user steers the MAV into an obstacle. Our system works when approaching 4 types of obstacles (well-textured, transparent, uniformly colored, poorly illuminated surface) from 6 different directions (Front, Back, Top, Bottom, Left, Right). In the third objective we demonstrate another part of the user interface, that allows to fix distance to 3 types of shapes: planes, cylinders (pipes), and girder elements (thin long objects). When fixing the distance to these surfaces user can only control MAV in the way that doesn't change the distance to the surface. Other degrees of freedom are controlled automatically by the MAV. In this objective we also evaluate the bandwidth required to operate the drone. In the fourth objective we demonstrate teaching MAV the view-poses and automatic revisiting them during the second autonomous flight. We evaluate translation and rotation errors between the poses of original and revisited view-poses and present the results of the inspection to the user as a list of associated images. This objective demonstrates a combination of visual localization and occupancy mapping with obstacle-avoiding trajectory planning for MAV.

8.3.1.1 Hardware and Environment Setup

Our freestyle has been carried out in the flying arena of the challenge host (ETH Zurich) on a modified structure mockup that has been used for benchmarking. To evaluate the second objective a transparent polycarbonate sheet of size 1×0.5 m has been attached to the mockup. For the third objective a pipe with diameter of 110 mm and length of 2 m and a wooden pillar that simulated a girder element were added to the scene.

We have used the MAV platform provided by the challenge host, that already comes with a Skybotix VI-sensor supporting 4 cameras with a wide field-of-view (ca. 122°), 2 of them in a stereo setup in frontal-downward direction. Additionally, an RGB-D camera (Asus Xtion) and 10 ultrasonic distance sensors (Devantech SRF02) were mounted on the drone to cover blind spots of the cameras and to increase the variety of perceivable obstacles (e.g. transparent, poorly illuminated surfaces). The MAV together with the sensors used in this task is shown in Fig. 8.2.

8.3.1.2 Results

In the first objective, the MAV has flown a predefined trajectory. It was using SLAM system to localize itself during the flight and simultaneously build the map of the environment. SLAM system maintained a keyframe map of the environment, where each keyframe had an associated depth image. After the drone has reached the end of the trajectory, an octree representation of the environment has been generated by projecting the depth measurements from all keyframes maintained by the SLAM

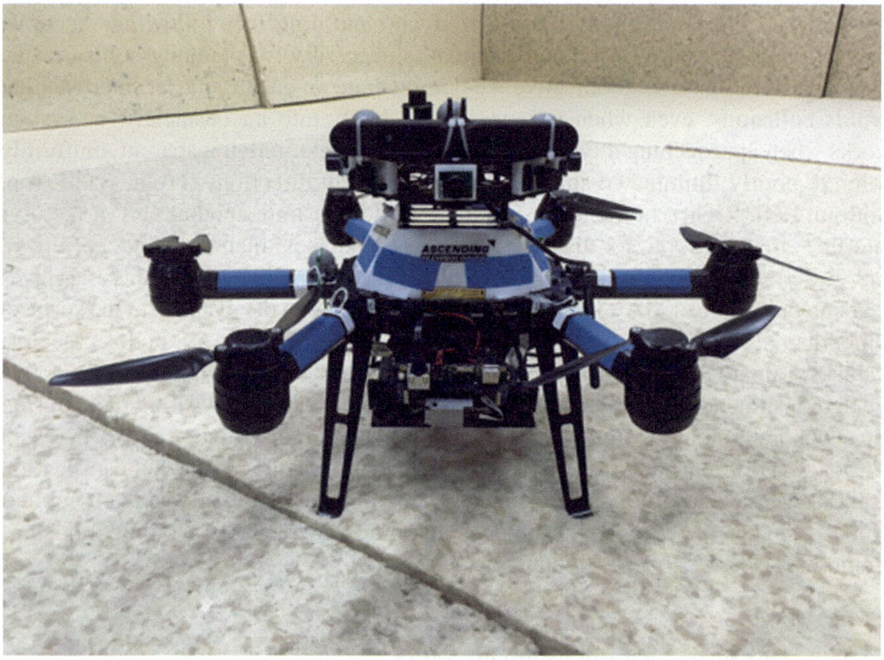

Fig. 8.2 MAV with RGB-D and ultrasonic distance sensors

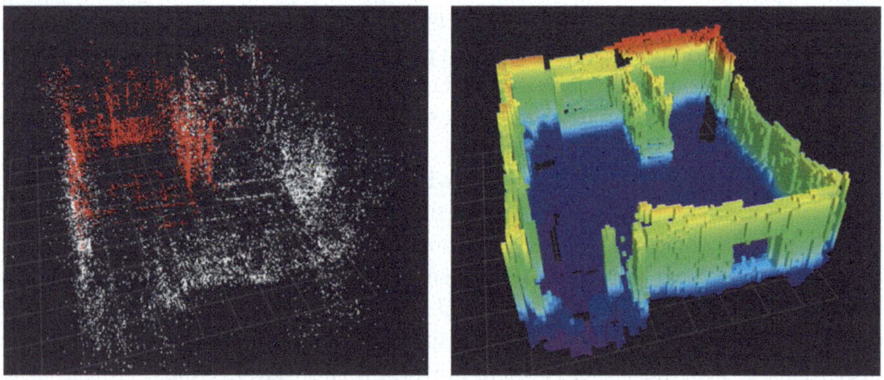

Fig. 8.3 Keypoint map generated by visual odometry and 3D occupancy map

system. The keypoint-based map used for localization and the octree map used for obstacle avoidance and trajectory planning are shown in Fig. 8.3.

When evaluating the second objective MAV has demonstrated successful collision avoidance for four types of obstacles (well-textured, transparent, uniformly colored, poorly illuminated surface) from six different directions (front, back, top, bottom, left, right). MAV was intentionally steered by the user to the obstacle,

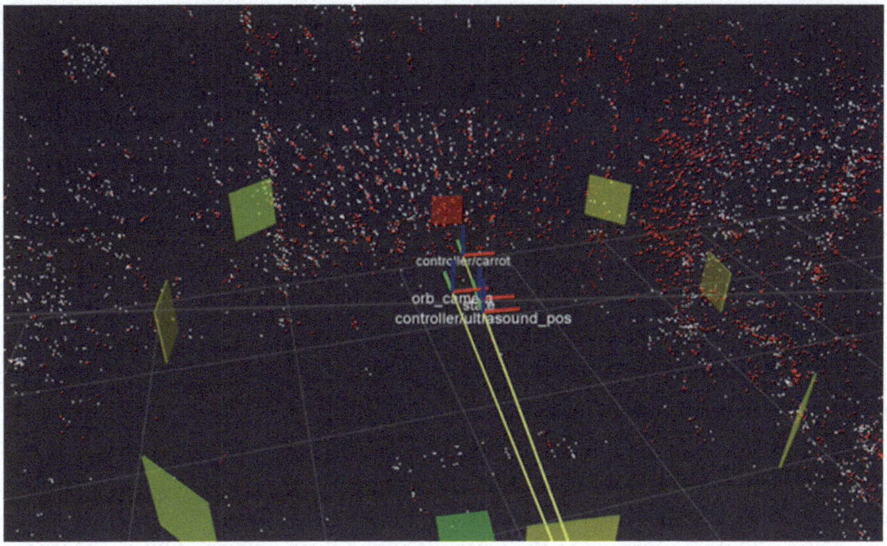

Fig. 8.4 Visualization of reactive collision avoidance from the left

but semi-autonomous interface could stop the MAV before the safety controller intervened. Figure 8.4 shows a visualization in the user interface when steering MAV to hit an obstacle on the left. Current position of the MAV is visualized as `state` coordinate frame. Desired position commanded by user is visualized as `controller/carrot` frame. Measurements from different ultrasound sensors are visualized with colors changing depending on a distance (green—far, red—close). Commanded position from collision avoidance system is visualized as `controller/ultrasound_pos` frame.

For the third objective the MAV demonstrated the ability to maintain a fixed user-defined distance to three different types of objects: cylinder (pipe), inclined plane and girder element. When fixing the distance all user commands that would lead to changing it are ignored, so the user can only move the MAV along the axis of cylinder and girder element, and parallel to the surface of the plane. This results in much easier user-control of the MAV close to these surfaces. During this flight a bandwidth between the MAV and ground station was evaluated, which was permanently below 1MB/s.

To evaluate the last objective, the following workflow has been used:

1. The operator starts the MAV from its starting zone.
2. The operator pilots the MAV to teach-in inspection waypoints using the semi-autonomous operator interface.
3. The operator pilots the MAV back to the starting zone. Trajectory to revisit the poses is planned.
4. During the second flight the MAV navigates autonomously to the learned inspection waypoints and collects new images.

5. The MAV navigates back to the starting zone autonomously. Results of two inspection flights are presented to the user. In our freestyle we have successfully revisited around 20 waypoints with average distance between original and revisited poses less than 10 cm and 5°.

Overall, in the freestyle task we have demonstrated the visual SLAM and 3D reconstructions on-board of the MAV, an assistive piloting mode that displays the 3D surrounding of the MAV and autonomously avoids collisions with obstacles from all directions, and is able to maintain fixed distance to different types of surfaces,and autonomous waypoint navigation to the learned inspection poses.

8.3.2 Showcase Stage

In the showcase task we have demonstrated systematic inspection in a medium-scale realistic indoor environment (flying arena provided at challenge host ETH Zurich). The core technologies shown in the freestyle round were enhanced by our team for operation at a larger scale in the new target environment. In addition, automatic image analysis for change detection has been demonstrated. The showcase task has demonstrated several features for the use-case of the end-user in a medium-scale indoor setting: the semi-autonomous operator interface, inspection mission planning interfaces, image analysis, and autonomous navigation by the MAV to inspection view poses.

A realistic medium-scale structure mockup, shown in Fig. 8.5, was constructed at the flying arena of the challenge host at ETH Zurich, and used for the showcase task. The dimensions of the mockup were 5/2/3 m (l/w/h).

In a first sub-task, we demonstrated the enhanced autonomous MAV navigation capabilities, adapted to the new requirements of the realistic industrial lab environment. One major challenge in this task was to scale vision-based localization and mapping, occupancy mapping, and path planning to the medium-scale environment.

To cope with the complexity of the environment we have used a two-layer visual pose estimation. The first layer runs real-time robust visual inertial odometry, which is used for controlling the MAV. It is shows limited precision, but runs in real time with limited computation resources and is robust to fast motions because of IMU integration. In the second layer we use a state-of-the-art SLAM method to localize the MAV more precisely. We have implemented map saving and loading functionality, which allows our system to re-localize itself in previously recorded maps. This method runs at lower frequencies and is used as localization source to correct for the accumulated drift over the flight. The second layer runs on the RGB-D data from the sensor that is mounted on the drone (Fig. 8.2), and due to the loop-closure capabilities is able to produce consistent maps. By saving depth maps for the corresponding keyframes we are able to reconstruct an occupancy map of the environment. The example of such map can be seen in Fig. 8.6.

Fig. 8.5 Mockup of the bridge constructed at the ETH Zurich for showcase demonstration

In this sub-task we also demonstrate the abilities of the semi-autonomous operator interface, which is able to maintain the fixed distance to the inspected surface. With this interface the operator is able to inspect the surfaces without direct line of sight, which was demonstrated during the showcase demonstration. During this sub-task we have also evaluated the trajectory and map accuracy metrics:

- Median Relative Pose Error on 5m segments (real-time): 0.414 m
- Median Relative Pose Error on 5m segments (localization): 0.099 m
- Median Absolute Trajectory Error (real-time): 1.007 m
- Median Absolute Trajectory Error (localization): 0.105 m
- Mathews correlation coefficient between computed and ground truth 3D obstacle maps: 0.827.

In the second sub-task, the MAV was able to autonomously navigate between view poses of a planned inspection mission. To define such an inspection mission, we have developed an operator interface for specifying missions on a 3D model of the structure. This model can be initially obtained through a mapping flight with the semi-autonomous flight mode. The interface with interactive waypoints is shown in Fig. 8.7.

In our showcase demonstration the system was able to conduct three autonomous flights through the same waypoints without triggering the safety controller. During the first flight, the MAV inspected the original structure. For the next flights artificial

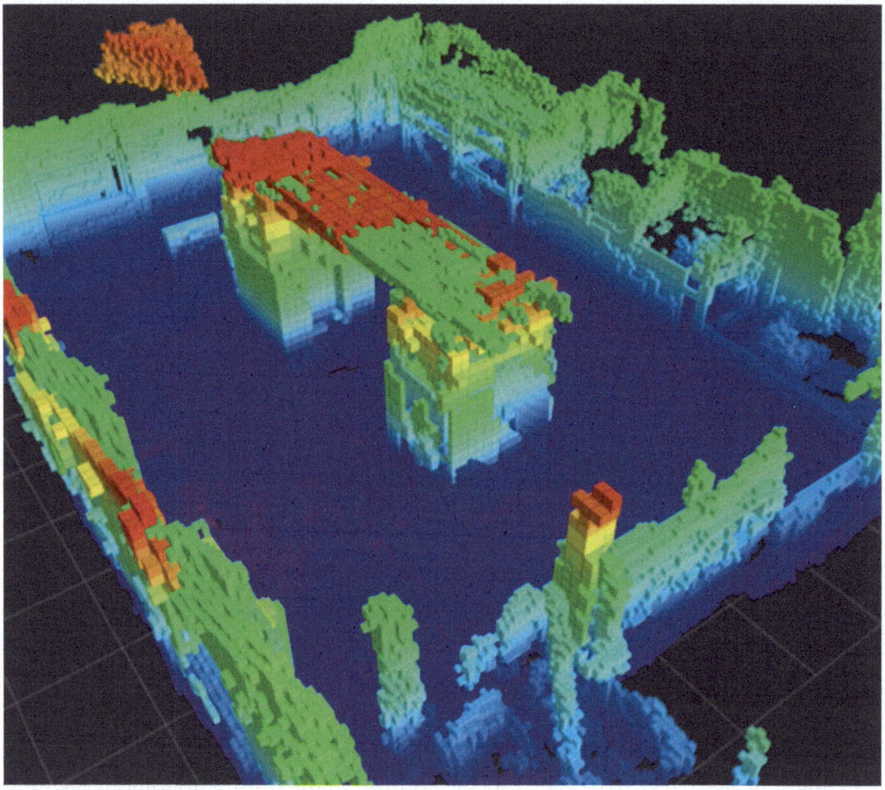

Fig. 8.6 Map built by the operator in the semi-autonomous mode overlaid with ground truth map

changes were introduced to the structure at different places. An example of the planned path is shown in Fig. 8.8. We were able to achieve an average error between inspected and re-inspected waypoints of 4.38° (angular) and 0.227 m (translation).

We have also developed an image analysis method for automatically detecting and ranking images for local surface changes on the structure. The aim is to assist inspection personnel in finding such changes in the images collected during inspection flights. In order to measure local surface changes in the images from similar view poses, we employ keypoint-based image registration and warping. Misalignments can be expected to occur where the surface has changed between inspections. An example of such change is shown in Fig. 8.9.

Our automatic system was able successfully detect 2 out of 2 different changes that were introduced to the environment. The full set of captured waypoints from 3 different flights, together with visualization of different steps of the change detection process is shown in Fig. 8.10.

To summarize, in our showcase we have demonstrated the following process:

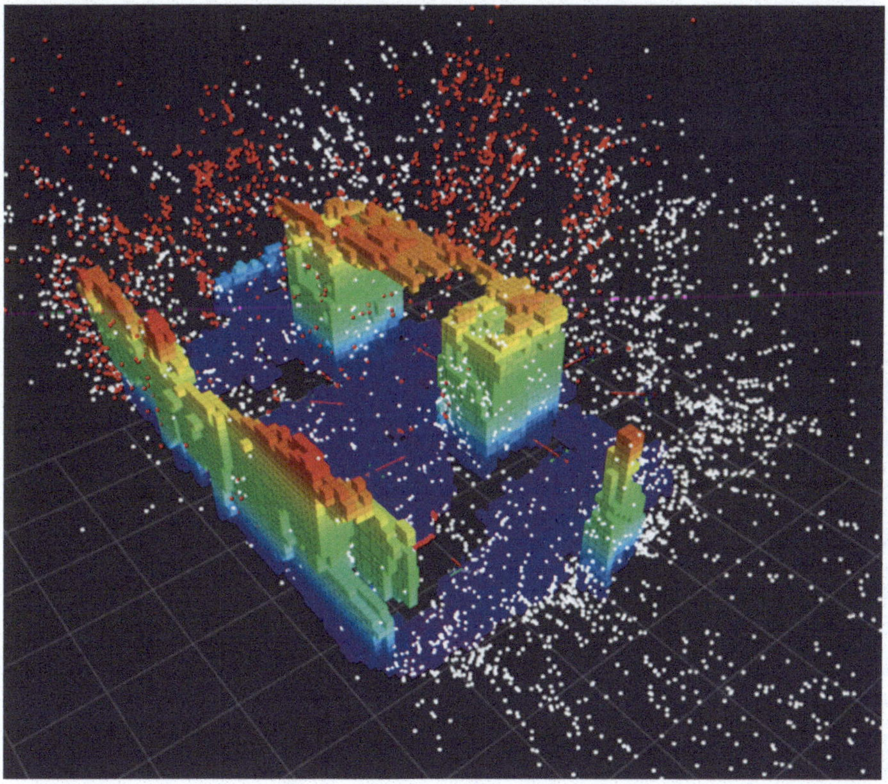

Fig. 8.7 An operator interface, where user can adjust the positions of the waypoints, visualized as red arrows

- The operator starts the MAV from its starting zone using the semi-autonomous operator interface.
- The operator pilots the MAV to teach-in a 3D map of the workspace using the semi-autonomous operator interface.
- The operator pilots the MAV back to the starting zone using the semi autonomous operator interface.
- The operator specifies an inspection mission on a 3D model.
- The MAV navigates autonomously to the inspection waypoints.
- The MAV navigates back to the starting zone autonomously.
- Local surface changes are applied to the mockup.
- The MAV reinspects the mockup and tries to detect the changes.
- Another set of local changes is applied to the mockup.
- The MAV reinspects the mockup and tries to detect the changes.

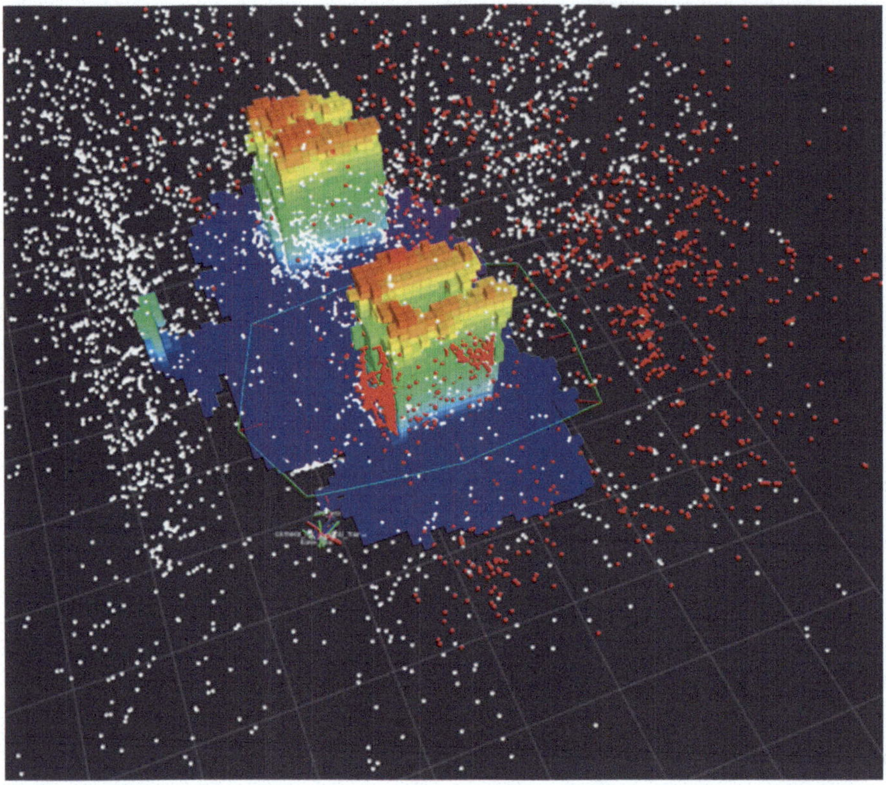

Fig. 8.8 Planned path through the user-defined waypoints

Fig. 8.9 Visualization of automatic change detection process. Color images from flight 1 and 2, warped image from flight 1 to the perspective of flight 2, and residuals are shown (left to right)

8.3.3 Field Test Stage

In the field test we have demonstrated systematic inspection of a bridge in a realistic outdoor environment. For the demonstration we selected a support column of the Sitterviadukt in St. Gallen, Switzerland (see Fig. 8.1). The core technologies shown in the showcase round were enhanced for operation at a larger scale in the new target environment. In addition, automatic image analysis for change detection has been

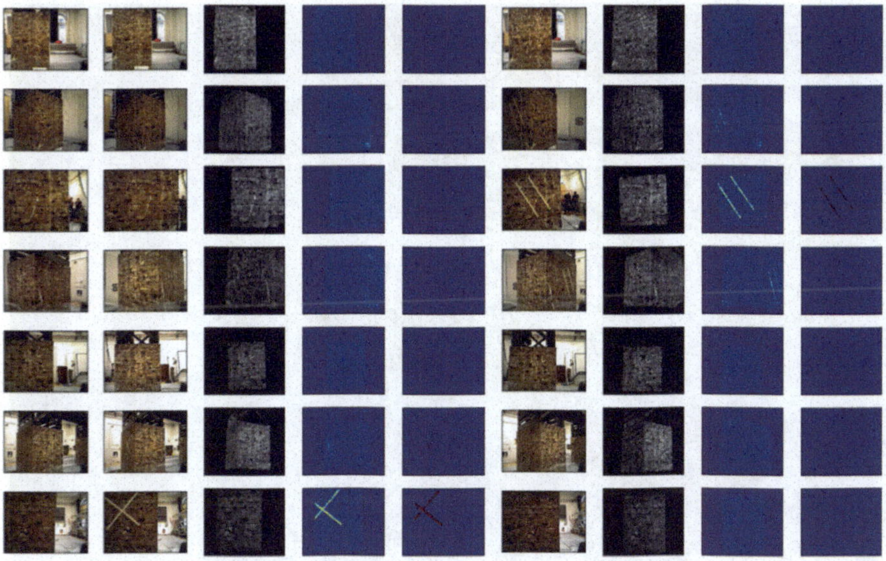

Fig. 8.10 Images collected from the waypoints at 3 different flights and visualization of the change detection algorithms. Each row represents one waypoint. The columns represent as color images from 3 flights, warped intensity images from flight 1 to the viewpoint of current flight, residuals after alignment and residuals with applied threshold

demonstrated. The field test has demonstrated several features for the use-case of the end-user in an outdoor setting: the semi-autonomous operator interface, inspection mission planning interfaces, image analysis, and autonomous navigation by the MAV to inspection view poses.

In the field tests, we demonstrated the enhanced autonomous MAV navigation capabilities, adapted to the new requirements of the outdoor environment. One major challenge was to adapt vision-based localization and mapping, occupancy mapping, and path planning to the outdoor environment. To cope with the complexity of the environment we have used a two-layer visual pose estimation running on the stereo data from the VI-Sensor that is mounted on the drone (Fig. 8.2). The first layer runs real-time robust visual-inertial odometry, which is used for controlling the MAV. It has limited precision, but runs in real time with limited computation resources and is robust to fast motions because of IMU integration. In the second layer we use a state-of-the-art SLAM method to localize the drone more precisely. We use map saving and loading functionality, which allows our system to re-localize itself in previously recorded maps. This method runs at lower frequencies and is used as localization source to correct for the accumulated drift over the flight. The second layer is able to produce consistent maps due to its loop-closure capabilities. By saving stereo images and computing depth maps for the corresponding keyframes we are able to reconstruct an occupancy map of the environment. An example of such a map can be seen in Fig. 8.11.

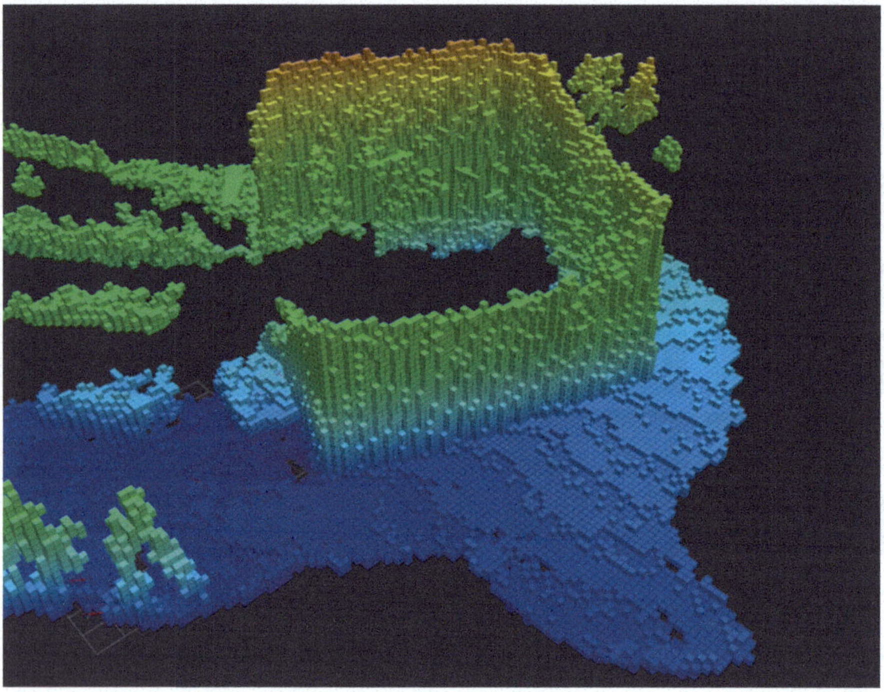

Fig. 8.11 Occupancy map acquired from the bridge structure

We also developed a semi-autonomous operator interface (Fig. 8.12) which implements position-control based on visual-inertial odometry and is able to maintain a fixed distance to the inspected surface. With this interface the operator is able to easily inspect the surfaces, which was demonstrated during the final field demonstration as a first sub-task. The vision-based position control of the MAV supports the operator in manually flying the MAV to positions in GPS-denied environments despite of wind. The interface uses the depth map from RGB-D sensor to detect the inspected surface. The surface is detected by fitting a plane using RANSAC into the pointcloud generated from the depth map. After detecting, the information about the plane is used by the controller to fix the distance to the plane and allow easier inspection by the human pilot.

In the second sub-task of the final field test demonstration, the MAV was able to autonomously navigate between view poses of a planned inspection mission. To define such an inspection mission, we have developed an operator interface for specifying missions on a 3D model of the structure. This model can be initially obtained through a mapping flight with the semi-autonomous flight mode. The interface with interactive waypoints is shown in Fig. 8.13.

In our final field test demonstration, the system was able to conduct four autonomous flights through two different sets of waypoints. During the first flight

Fig. 8.12 The system maintains a fixed distance to the inspected surface in semi-autonomous mode

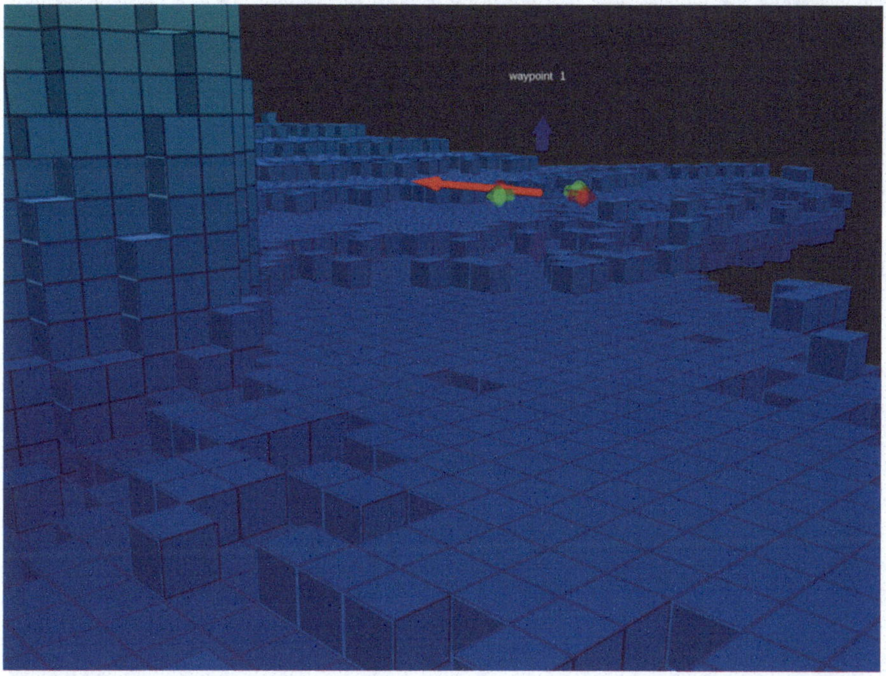

Fig. 8.13 An operator interface in which the user can adjust the positions of the waypoints, visualized as red arrows

of each trial, the MAV inspected the original structure. For the next flights artificial changes were introduced to the structure at different places. An example of the planned path is shown in Fig. 8.14. For the first trial, the operator piloted the MAV to acquire a new map of the workspace and defined new inspection mission waypoints. In the second trial, a pre-built map and inspection mission from a previous field test several weeks ago has been used.

We have also developed an image analysis method for automatically detecting and ranking images for local surface changes on the structure. The aim is to assist inspection personnel in finding such changes in the images collected during inspec-

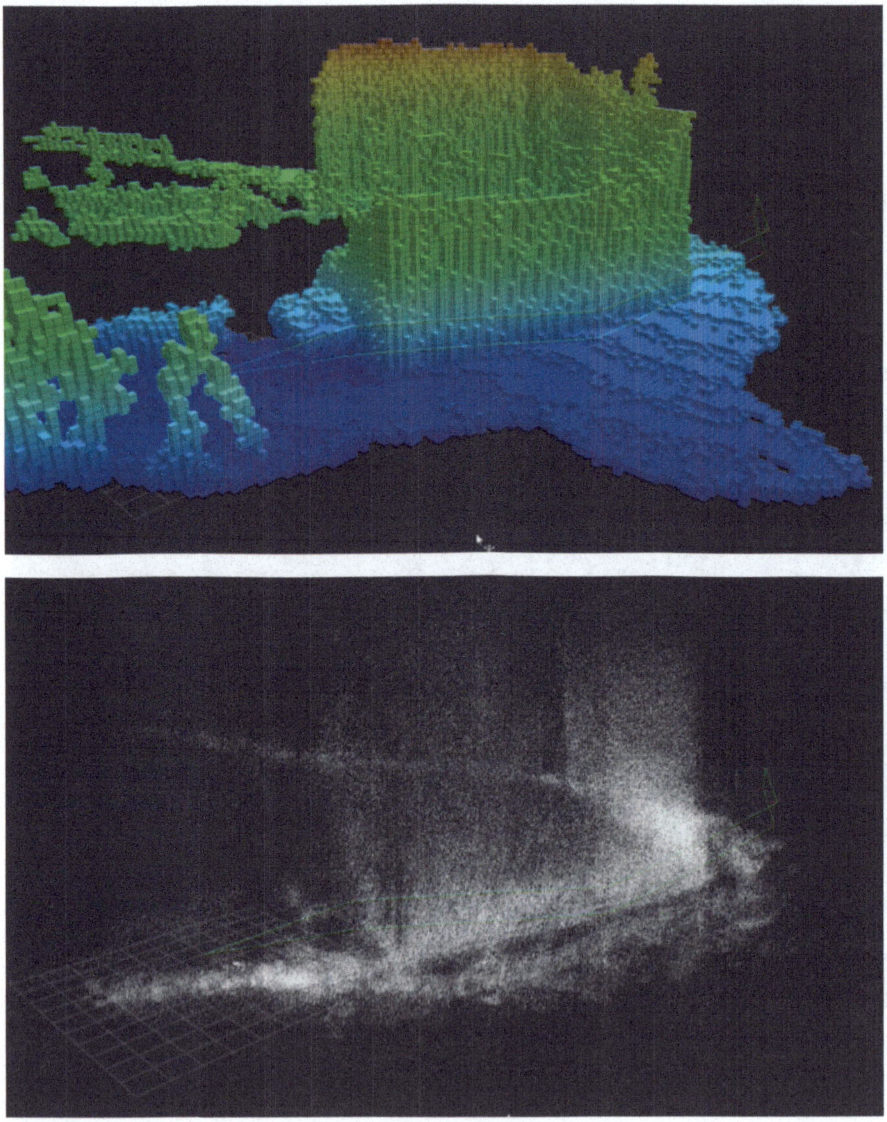

Fig. 8.14 Planned path through the user-defined waypoints

tion flights. In order to measure local surface changes in images from similar view poses, we employ keypoint-based image registration and warping. Misalignment can be expected to occur where the surface has changed between inspections. An example of such a change is shown in Fig. 8.15.

Our automatic system was able to successfully detect all changes in two trials that were introduced to the environment with one false positive in the second trial. Due

Fig. 8.15 Visualization of automatic change detection process. Color images from flight 1 and 2, warped image from flight 2 to the perspective of flight 1, and residuals are shown (left to right, top to bottom)

to the bright and sunny lighting conditions on the test day, the spinning rotors were visible in the images due to small exposure time. False detections in the image parts containing the rotors have been avoided by cropping the images accordingly. The full set of captured images from the second trial during the field tests, together with visualization of different steps of the change detection process is shown in Fig. 8.16.

To summarise, in our final field test demonstration we have shown a semi-autonomous control interface that keeps a fixed distance to the wall of the building structure for inspection. We also demonstrated the following process for automated inspection missions:

1. The operator starts the MAV from its starting zone using the semi-autonomous operator interface (vision-based position control).
2. The operator pilots the MAV to teach-in a 3D map of the workspace using the semi-autonomous operator interface (vision-based position control).
3. The operator pilots the MAV back to the starting zone using the semi-autonomous operator interface (vision-based position control).
4. The operator specifies an inspection mission on a 3D model of the bridge structure.
5. The MAV navigates autonomously to the inspection waypoints and takes images.
6. The MAV navigates back to the starting zone autonomously.
7. Local surface changes are applied to the bridge structure.
8. The MAV autonomously reinspects the bridge structure and tries to detect the changes (using the map and inspection mission from steps 1–4).

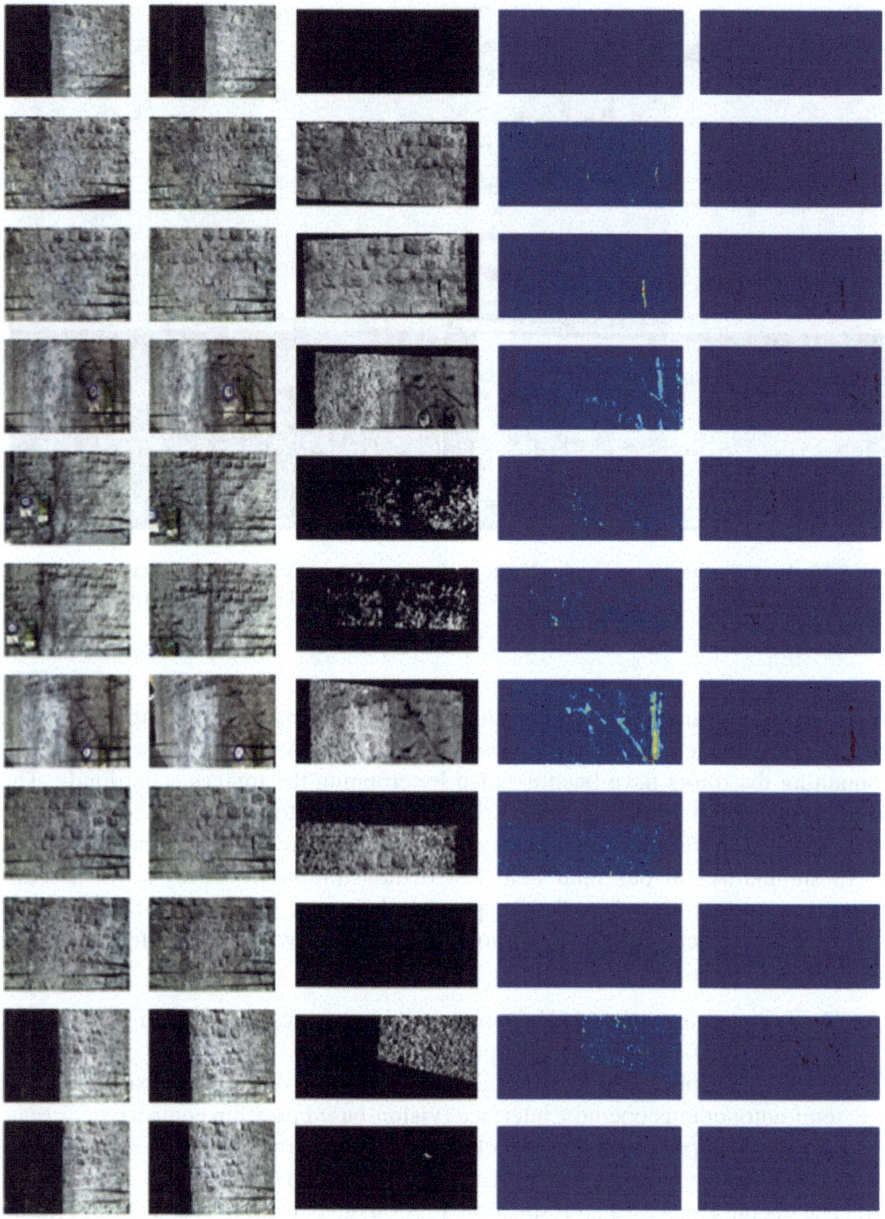

Fig. 8.16 Images collected from the waypoints during the second trial in 2 flights and visualization of the change detection algorithm. Each row represents one waypoint. The columns represent colour images from 2 flights, warped intensity images from flight 2 to the viewpoint of flight 1, residuals after alignment and residuals with applied threshold

9. The MAV navigates autonomously to a second set of inspection waypoints (using a pre-built map and inspection mission from a previous field-test) and takes images.
10. The MAV navigates back to the starting zone autonomously.
11. Another set of local changes is applied to the bridge structure.
12. The MAV autonomously reinspects the bridge structure and tries to detect the changes (using the pre-built map and inspection mission from the previous field-test).

Before the final demonstration of the field test a series of outdoor tests in Garching, Germany (TUM campus) and St. Gallen, Switzerland were conducted. During the tests some issues with spare parts for the MAV hardware were identified. For the final demonstration the issues were resolved by AscTec through replacing the MAV with a newer version (11-inch version of AscTec Neo). The low temperature at the test day (around $-10\,°\mathrm{C}$) required preheating of the batteries and sensors. To the best of our knowledge, there are no drones on the market that can localize on the map build from different flight based on vision and systematically revisit the user defined waypoints.

According to the defined objectives, we demonstrated the feasibility of the use-case of systematically inspecting structures of a bridge in challenging outdoor conditions (temperature around $-10°$ Celsius, changing light conditions, wind, no safety controller). During the demonstration the MAV performed real-time vision-based localization, 3D reconstruction, obstacle-avoiding flight, autonomous navigation to systematically inspect a structure and image analysis for local surface change detection. We also demonstrated a semi-autonomous operator interface that is able to maintain a fixed distance to the inspected surface.

CVG' s task in the field test round was the development of advanced software technologies for the MAV inspection system, including vision-based navigation, autonomous and semi-autonomous flight modes, inspection planning, and automatic image analysis for changes. CVG has also integrated its software on the MAV platform to implement demonstrations of the inspection system. CVG has used its extensive background in methods for vision-based localization and mapping, and vision-based navigation for MAVs.

SLB' s role in the team was to provide insights and definitions for the use-case of bridge inspection through its perspective of a service provider. As a provider of engineering and architectural solutions in the construction industry, SLB maintains a lot of contacts to possible customers and knows the requirements of inspection/survey tasks out of the practice. Since 2012, SLB provides MAV survey and inspection as a new service to its customers. In this time, SLB gathered a lot of knowledge about the current state-of-the-art of the MAV technology and its limitations. As a potential end-user and with the experience related to the working environment, SLB has brought the user perspective into the team. The close distance between the home bases of all team members and the demonstration locations facilitated the communication and exchange in the team as well as meetings for joint tests of the developed system and technologies. SLB has also organized access to the bridge and provided necessary equipment for the bridge inspection that was demonstrated during the field tests.

Our team has brought together CVG as an academic research partner and SLB as an industrial service provider and, hence, served the main objective of EuRoC to strengthen collaboration and cross-fertilisation between the industrial and the research community. SLB has shared its knowledge about real use-case requirements with CVG, which was used by CVG for the development of sustainable solutions that could be applicable to SLB' s use-case, other end-users, tasks and situations.

8.4 Conclusion

The results of this project demonstrated a set of technologies for autonomous bridge inspection with the MAVs and allowed both computer vision group of TU Munich and the industry partners to improve their relative position towards competitors. The computer vision group has tackled highly challenging research problems in computer vision and robotics. The activities on real-time vision-based SLAM, vision-based autonomous navigation, change detection, and image classification have generated a strong impact in both scientific communities and will sustain the position of the industry partners at the scientific frontier in these areas.

The pursued technologies have rich potential for further development of novel inspection service products. As an end-user of such future services, SLB could lead the way to MAV inspection services, expand the range of offered services, and strengthen its position on the market as inspection service provider. SLB was also able to raise the awareness level of the company and its MAV services through the participation in this project.

The employment of MAV technology for the inspection of structures such as industrial and power plants, or traffic infrastructure is highly beneficial for the EU economic growth. Maintenance is an essential activity for sustaining the economy. While conventional inspections impose large costs (climbing teams, access vehicles, long down-times), systematic mobile inspection with MAV technology would significantly reduce inspection time, increase repeatability, and cut down costs for inspections (about 30–50%). In addition, the number of potential objects continues to grow. This is especially the case for SLB' s use-case of bridge inspection (Table 8.1).

The work done in this project contributes to the improvement of the efficiency, adaptability, and sustainability of MAV applications and services in Europe. Through this EuRoC project, new MAV technologies have been customised, tested and validated in a real use-case scenario. By this, the aim is at improving their potential for applications and services which can compete on the global market.

Beyond bridge inspection, the advanced robotics technologies pursued in this project have large potential for innovations in further domains in services and industry such as automated logistics, autonomous cars, or service robotics. Such innovations will lead to novel high quality products and services that strengthen the global competitiveness of EU industry and SMEs and will open the markets of the future.

Table 8.1 Gives an overview of approximate number of bridges in different countries. Bridges should be examined in regular intervals of e.g. 3 years, from which approximately 30–40 could have a size relevant to MAV inspections. This means that alone in the EU, the number of bridges to be inspected per year is immense

Country	Road Bridges	Railway Bridges	Total	Remarks
Germany	110000	27000	140000	Replacement value approx. 80 bil Euro
Switzerland	10000	6100	16000	Replacement value approx, 19 bil CHF
EU			217000	
USA			607000	Probably only road bridges

The research done by CVG on vision-based localization, 3D reconstruction, and robot navigation will also support the EU's forefront scientific position in robotics and computer vision.

References

1. Cover, H., Choudhury, S., Scherer, S., Singh, S.: Sparse tangential network (spartan): motion planning for micro aerial vehicles. In: 2013 IEEE International Conference on Robotics and Automation, pp. 2820–2825 (2013)
2. Achtelik, M.W., Lynen, S., Weiss, S., Chli, M., Siegwart, Roland: Motion-and uncertainty-aware path planning for micro aerial vehicles. J. Field Robot. **31**(4), 676–698 (2014)
3. Abdel-Qader, I., Abudayyeh, O., Kelly, Michael E.: Analysis of edge-detection techniques for crack identification in bridges. J. Comput. Civil Eng. **17**(4), 255–263 (2003)
4. Chanda, S., Bu, G., Guan, H., Jo, J., Pal, U., Loo, Y.-C., Blumenstein, M.: Automatic bridge crack detection—A texture analysis-based approach. In: Gayar, N.E., Schwenker, F., Suen, C. (eds.) Artificial Neural Networks in Pattern Recognition, pp. 193–203. Springer International Publishing, Cham (2014)
5. Tian, J., Cui, S., Reinartz, P.: Building change detection based on satellite stereo imagery and digital surface models. IEEE Trans. Geosci. Remote Sens. **52**(1), 406–417 (2014)

Chapter 9
GRVC-CATEC: Aerial Robot Co-worker in Plant Servicing (ARCOW)

Jose Ramiro Martínez de-Dios, Francisco J. Pérez-Grau, Arturo Torres-González, J. Joaquín Acevedo, Julio L. Paneque, Antidio Viguria, Dimas Fuego, Juan R. Astorga and Anibal Ollero

Abstract The needs of higher levels of automation in the industrial processes and manufacturing have led to the paradigm of robot co-worker. Aerial robots provide interesting advantages when compared to traditional ground robots and can contribute to factory automation due to their intrinsic flexibility and fast response. For instance, the upper part in factories are rather sparse and static whereas the floor of factories are often very dense and dynamic. This chapter presents the ARCOW project in which

---

J. R. M. de-Dios (✉)
Robotics Laboratory, Universidad de Sevilla,
Avenida de los Descubrimientos s/n, 41092 Seville, Spain
e-mail: jdedios@us.es

A. Torres-González · J. L. Paneque · D. Fuego · A. Ollero
Robotics Lab Sevilla (Universidad de Sevilla) Avda. Camino de los Descubrimientos, Seville, Spain
e-mail: arturotorres@us.es

J. L. Paneque
e-mail: jlpaneque@us.es

D. Fuego
e-mail: dimasfuego@hotmail.com

A. Ollero
e-mail: aollero@us.es

F. J. Pérez-Grau · J. J. Acevedo · A. Viguria
CATEC (Advanced Center for Aerospace Technologies),
C/ Wilbur y Orville Wright 19, 41309 La Rinconada, Seville, Spain
e-mail: fjperez@catec.aero

J. J. Acevedo
e-mail: jose.joaquin6@gmail.com

A. Viguria
e-mail: aviguria@catec.aero

J. R. Astorga
Airbus Defence and Space, Parque Tecnológico TecnoBahía,
El Puerto de Santa María, Cádiz, Spain
e-mail: juan.r.astorga@airbus.com

F. Caccavale et al. (eds.), *Bringing Innovative Robotic Technologies from Research Labs to Industrial End-users*, Springer Tracts in Advanced Robotics 136,
https://doi.org/10.1007/978-3-030-34507-5_9

the University of Seville, FADA-CATEC and Airbus D&S joined efforts to develop and demonstrate how aerial robots collaborating with humans can be introduced in manufacturing processes. ARCOW is focused on two use-cases: light item logistics and localization of missing tools. This chapter describes the approaches, architectures and methods developed and validated in the different stages of EuRoC to accomplish this objective. The ARCOW system was extensively validated in the Airbus D&S Centro Bahía de Cádiz (CBC) factory in Spanish province of Cádiz and received the "Best drone-based solution" in the EU Drone Awards 2017.

Keywords Aerial robots · Robotic co-worker · Logistics · HMI · Localization · Navigation

9.1 Introduction

The needs of higher levels of automation in the industry together with new advances of robotic technologies have led to the paradigm of *Robot Co-Worker*. The greater majority of the robot co-workers in industrial sectors that have been proposed refer mainly to manipulators jointly collaborating with human operators in complex manufacturing processes, ground robots helping humans in transportation or exoskeletons [1, 2]. In this chapter we describe our participation in ARCOW (Aerial Robot Co-Worker in Plant Servicing) in which we research in the design, implementation and validation of aerial robots as co-workers in manufacturing in the aeronautical industry.

The new manufacturing paradigm in the aerospace sector pursues the acceleration of delivery rates through the gradual implementation of automated processes. In Europe this trend is driven by Airbus Group, parent company of Airbus Defence and Space (D&S), as one of the two major aircraft manufacturers in the world. Aerial robots can contribute to this automation process by providing intrinsic capabilities such as flexibility, fast response and availability, but at the cost of solving the challenge of safe indoor navigation in man-inhabited GPS-denied environments.

ARCOW is focused on the implementation of two particular challenging applications that will improve productivity by reducing operation times and manufacturing costs in aeronautics.

The first one is the logistic process of light goods, which involve hundreds of small parts required in aircraft manufacturing such as tools, bags of rivets, seals, cleaning napkins, among others. The delivery of these goods traditionally generates many displacements of the workers between the storage facilities and the assembly positions, which are usually performed at a significant height. The second application is the localization and identification of tools used by workers in manufacturing. Aeronautic industrial processes involve the use of hundreds of different specific tools. The loss of tools can cause delays in the production line and can cause Foreign Object Debris (FOD). Both applications have a large impact on the manufacturing operations: it is estimated that this can represent between 10 and 15% of the processing time, and

10% of the production lead time. The reduction of these waste times by using aerial robots can generate an important profit.

This chapter presents the ARCOW project in which the University of Seville, FADA-CATEC and Airbus D&S joined efforts to develop and demonstrate how aerial robots collaborating with humans can be introduced in manufacturing processes in order to reduce their costs and improve their efficiency. This chapter describes the approaches, architectures and methods developed and validated in the different stages of EuRoC to accomplish this objective.

The chapter is structured as follows. Section 9.2.1 describes the motivation of ARCOW and the selected use-cases. Section 9.2.2 briefly summarizes the main related work in the methods developed for ARCOW. Section 9.2.3 summarizes the scenarios where ARCOW developments and missions have been successively experimented while increasing their level of realism. Section 9.3.1 presents the architecture of the ARCOW system, defining the main functional modules and how they are related. In Sect. 9.3.2, the localization estimation techniques are summarized. It is divided in two parts, one describing the robot pose estimation method, and the other describing the tool localization method. The developed motion planning methods developed are summarized in Sect. 9.3.3. Finally, Sect. 9.4 summarizes the conclusions.

9.2 Use Case Description and Motivation

9.2.1 Motivation

ARCOW had a direct application-driven approach. There is a clear need in the aeronautical industry to increase the level of automation, which is currently low when compared to other manufacturing industries, by maintaining or increasing the aeronautical manufacturing quality and safety standards. Our approach is to integrate aerial robots collaborating with humans in industrial manufacturing processes. In these applications, aerial robots provide interesting advantages when compared to traditional ground robots:

- The floor of factories are often very dense and dynamic environments (e.g. objects and workers) for ground robot motion, whereas the upper part in factories is rather sparse and static, more suitable for aerial robot navigation.
- Safety constraints impose low-speed motions for ground robots, while this constraint is not present for aerial robots.
- Unmanned aerial systems (UAS) regulation does not apply in these confined environments.

Of course, introducing aerial robots in manufacturing processes poses interesting challenges that should be addressed in the project. From a technical perspective, the system should include GNSS-denied precise long-term robot pose estimation, as well as efficient planning methods suitable for tight and dynamic environments. From operational and implementation perspectives, the system should ensure safe aerial robot navigation in tight and crowded environments. Although ARCOW is conceived as a pilot project, and not as a product, we paid special interest to safety mainly in terms of the robustness of the methods developed for navigation.

Airbus D&S established two main use cases in ARCOW due to their relevance in the aeronautical manufacturing sector:

- Use of aerial robots to improve operational efficiency and reduce costs in logistic operations involving light objects.
- Use of aerial robots to improve operational efficiency in detection and localization of missing tools.

9.2.1.1 Light-Object Logistics

The assembly of aerostructure parts requires many different materials and small and light components (rivets, sealing materials, glues, etc). For instance, in the Airbus D&S factory where EuRoC Field Tests were performed, more than 500 different small components and materials are necessary for the assembly of every single fan cowl. The transportation of light components (<500g) constitutes around 98% of the logistic operations in this factory, which yields the importance of reducing costs in such operations. Moreover, a number of these components such as sealants are kept refrigerated before their use, and they must be heated just before using them, hence limiting their lifetime. These constraints together with the high manufacturing quality requirements of aeronautical industries, and the very wide variety of small components, make light object logistics very relevant in the assembly of aerostructure parts.

All these components are located at specific storage centers, at different places in the factory. In the current Airbus D&S operation, when a worker runs out of some component, he/she informs his/her supervisor who starts a procedure to ask for it. This procedure ensures traceability and quality requirements. When the component is ready, the supervisor brings the component to the worker.

The objective is to use aerial robots for light-object logistics within the factory. The aerial robot is in charge of transporting and providing light components to the workers that demand them. For the transportation, the light components are inserted in capsules of a suitable size. The aerial robots are equipped with a cargo mechanism for the transportation and deployment of these capsules. When a worker needs a component, he informs the ARCOW system using a Human-Machine Interface (HMI), e.g. a tablet running ARCOW application. This request is sent to the storage center, where the component is then prepared and placed in the cargo mechanism, and then the aerial robot is commanded to perform a transportation mission to a specific delivery point in the factory. Finally, when the component is delivered, the worker acknowledges reception through the same HMI. ARCOW system also ensures Airbus D&S traceability and quality requirements.

9.2.1.2 Localization of Missing Tools

The assembly of aerostructure parts requires many different tools. For instance, in the Airbus D&S factory where EuRoC Field Tests were performed, more than 100 different tools are necessary for the assembly of every single fan cowl. These tools are very specific. Missing tools is a problem for two reasons: (1) it may cause delays and reduce the productivity of the assembly line, and (2) missing items may cause FOD.

Currently, the factory is equipped with intelligent tool closets, which detect when a specific tool is missing at the end of the working shift. When a tool is considered missing, it is searched in the factory requiring significant human labour. ARCOW project implemented and tested a functionality where the aerial robots search and locate these missing tools. The aerial robot is equipped with a radio emitter than can send requests and receive responses from missing tools. Each tool is equipped with a radio tag with an ID that can listen to and respond to request messages.

The operation of this use case in ARCOW is as follows. When a tool is considered missing, its ID is inserted into the list of the tools that must be searched. During the next flights, the aerial robot transmits request messages with the IDs of the list of missing tools. When the radio tag of a missing tool receives a request with its ID, it responds. The aerial robot receives the responses and estimates the location of the mission tools within the factory map. The location of the missing tool is shown in an HMI located at the storage center. The system can search simultaneously many missing tools. It is also possible to perform missing tool search flights on-demand, which are interesting for instance to urgently locate specific missing tools. If at the end of the working shift a tool is considered missing, the IDs of the missing tools are also inserted in the list, and a search flight is performed on-demand.

9.2.1.3 Expected Impact

Airbus D&S is very interested in the introduction of automation in the strategy of Airbus Group regarding the factory of the future. The increasingly competitive aircraft market leads to the introduction of innovative technologies in the manufacturing process in order to increase productivity. Airbus Group, as parent company of Airbus D&S, has 144.061 employees and, with around 60 billion euro revenues yearly, is the biggest aeronautics company in Europe and the main Boeing competitor in the world aeronautics industrial sector. Not only the development of innovative final products contributes to the positioning in the aeronautical sector, but also the capability to manufacture with reduced delivery times and manufacturing costs.

Logistic operations have an important impact on the total cost of assembly and manufacturing operations in Airbus. For example, in the factory where EuRoC Field Tests were performed, there is a group of 15 people exclusively devoted to internal logistic operations. The implementation of the aerial robot technology for light-object logistics is estimated to have the following direct savings in an aeronautical manufacturing plant:

- Reduction in logistic costs: the use of aerial robots could allow reducing the number of people devoted to logistics operations.
- Reduction in lead times: the use of aerial robots reduces the non-productivity time of the workers and allows producing faster than with classical logistic systems.

The improvements in the manufacturing process of the light-object logistic use case were classified into: (a) savings in logistics, and (b) savings in lead time and recurring costs.

On the other hand, and in relation to the second use case of ARCOW system, another key objective of aircraft manufacturing plants is to increase the quality of the processes through innovation, which has an important added value for Airbus clients. It was estimated that the implementation of the missing tools search system is expected to significantly decrease the time devoted to find lost objects in assembly operations, thus decreasing wasted time in assembly and manufacturing operations. Also, it will improve the trust of the clients by reducing FOD incidents. Hence, aerial robots will be used to increase the quality of manufacturing processes by reducing the time to find missing objects or tools. Although this benefit is difficult to quantify, the increment of quality of the processes has great importance and value in Airbus manufacturing plants.

9.2.2 Related Work

For brevity, we focus the summary of the related work on robot localization and navigation, which are basic pillars of mobile robotics. They have been extensively researched in the last three decades for ground robots with great results in both GPS-enabled and GPS-denied areas.

Although not in all cases, 2D navigation and localization can be assumed in most ground robots. Extending these algorithms to 3D poses serious challenges that go further than space dimensionality. In most cases, aerial robots cannot assume the existence of reliable odometry systems, as it is usually assumed in ground robots. Many works rely on fusing inertial sensors and altitude sensors (such as lasers or barometers) with monocular vision in order to estimate the robot localization based on visual Simultaneous Localization And Mapping (SLAM) [3–5]. These approaches work very well when the same area is repeatedly visited, but use to fail at high-speed aerial robot motions in unknown scenarios. Optical flow approaches [6, 7] make use of the same sensors but aim to estimate only linear velocities. These approaches have demonstrated to work very well, however velocity integration for localization quickly diverges. Then, these methods cannot be used for long-term localization.

Stereovision systems provide direct depth measurements in a 3D environment. SLAM [8] and odometry approaches [9] have shown good results at short-term localization and even in long trajectories without loop closing. Many of the algorithms developed for stereovision odometry and SLAM can be applied to 3D cameras based on pattern projection [10, 11]. Also, point-cloud based odometry and localization is another significant research area. These methods integrate geometric point-clouds of the environment usually provided with LIDAR or Time of Flight (ToF) cameras. The methods applied for velocity or position estimation compute the alignment between two consecutive point-clouds or between the registered point-cloud and a map representing the environment, using the classic Iterated Closed Points (ICP) [12] or variations of it. Although computationally expensive, recent advances in Kd-Trees [13] and Approximate Nearest Neighbours (ANN) [14] enable fast computation in the presence of medium and dense point-clouds.

Different approaches can be used for fast and reliable robot navigation: standalone global navigation, global plus local navigation or just relative navigation. The approach selected for ARCOW takes into account that the robot navigation should consider high-level information, such as the integration with other systems, avoiding crowded areas or following of specific paths to avoid critical zones. Hence, a global+local navigation scheme will be used as in [15]. The key idea is to build a graph connecting the different paths in the environment so that the global planner can compute the shortest path through the graph from the current location to the goal position. The local planner will execute the plan locally dealing with non-modelled and/or dynamic obstacles. Many planning methods have been developed and tested. Rapidly-exploring Random Tree (RRT) approaches—such as [16, 17]—are flexible randomized schemes that provide any-time solutions with low computational resources. Theta* methods, based on A*, are also good alternatives to find realistic and short-looking paths.

Finally, radio tags will be integrated into ARCOW for the localization of the aerial robot and the missing tools. These devices exhibit low weight, size and power consumption. Hence, they fit perfectly with wearable devices or aerial robots. The main drawback of point-to-point distance-based localization schemes is a low informative measurement that leads to multi-hypotheses localization. ARCOW applied latest state of the art algorithms to solve the radio tag localization problem. A multi-hypotheses approach based on the combination of Particle Filters (PFs) and Kalman Filters (KF) has been used for simultaneous localization and mapping of radio tags and the aerial robot (called Range-Only SLAM or RO-SLAM), see e.g. [18, 19].

9.2.3 Scenarios

This section summarizes the scenarios in which the ARCOW system was incrementally developed and validated during EuRoC. EuRoC Stage 2 was performed in two different testbeds: one at ETH Zurich (Stage2.a) and another one in CATEC (Stage2.b). EuRoC Stage 3 was performed in one of Airbus D&S factories.

Fig. 9.1 The aerial robot performing obstacle detection and avoidance of a human worker represented by a mannequin (left) and another smaller aerial robot (right)

9.2.3.1 Stage2.a Scenario: ETH's Testbed

EuRoC Stage 2.a consisted of performing a series of tasks defined by the Challenge Host (ETH Zurich), which were the same for all the EuRoC teams competing in Challenge 3, using the same flying arena with the same base platform.

Besides the *Benchmarking* experiments, EuRoC Stage 2.a also included the so-called *Free-Style* tests, that were chosen by each team in order to validate specific functionalities according to their project use cases. In our case, we performed experiments related to the introduction of aerial robots in aircraft manufacturing plants. These experiments included tasks related to the interaction of aerial robots with other agents in their environment, such as human workers or other aerial robots, as well as the localization of missing tools using radio sensing. Hence, obstacle detection and avoidance were basic tasks that our system needed to tackle. Figure 9.1 shows two pictures taken during the *Free-Style* experiments performed in ETH's flying arena: obstacle detection and avoidance of workers using a mannequin (left) and of other aerial robot (right).

ETH's testbed is a closed room with a size of $9 \times 7 \times 2$ m [20] equipped with a VICON motion capture system that was only used for aerial robot tracking and obtaining evaluation metrics. The testbed also contains scaffold elements at the walls which can be moved throughout the arena. The structure consists of thin poles which set high demands in terms of perception accuracy and robustness, as well as collision-free navigation capability. Other elements used as obstacles in the room were a rusty boiler and two wooden pallets.

For both the *Benchmarking* and the *Free-Style* rounds, our team used the aerial platform that was provided by the Challenge Host, a Neo UAS from Ascending Technologies equipped with a visual-inertial sensor for localization and obstacle detection. The only modification we introduced was the integration of an Ultra-Wide Band (UWB) receiver as the on-board radio sensor, as depicted in Fig. 9.2.

Fig. 9.2 The aerial robot used during Stage2.a

9.2.3.2 Stage2.b Scenario: CATEC's Testbed

The *Showcase* round paved the way to implement the planned use-case driven tasks as part of Stage 2.b. This round was validated in a controlled environment emulating the scenario of an aircraft manufacturing plant or aircraft assembly line. This scenario was replicated in the indoor testbed of CATEC facilities, where industrial scenarios have been already replicated in other projects. Figure 9.3 shows the aircraft manufacturing environment replicated in CATEC's testbed.

CATEC's indoor testbed is used to develop and test a wide range of technologies related to autonomous systems. The useful volume where tests can be conducted is a cube of 15x15x5 meters, which allowed for a realistic representation of the end user scenario. The testbed houses an indoor localization system based on 20 VICON cameras. This system is able to provide the position and attitude of each object in real-time with millimetre precision, and with very low latency in communications. The aerial robot localization data was used for evaluation purposes only.

9.2.3.3 Stage3 Scenario: Airbus D&S CBC Factory

The Airbus *Centro Bahía de Cádiz* (CBC) factory is an industrial center located in El Puerto de Santa María, in the Spanish province of Cádiz. It is an international referent in the aeronautical industry and builds products for airplanes such as the Airbus A400M, A300MRRT, A320-Neo and Boeing 737MAX, among others.

The use case scenario of ARCOW is the fan cowl manufacturing plant at CBC, see Fig. 9.4. Fan cowls (can be seen in clear blue in the picture) are the protective

Fig. 9.3 Replicated environment at CATEC indoor testbed

Fig. 9.4 View of the Airbus D&S CBC fan cowl manufacturing plant

elements that cover the engines of airplanes. They are one of the most important components of an airplane's structure, and the CBC factory is one of the global leading fan cowls manufacturers.

The distribution of the manufacturing plant is complex and dynamic: fan cowls are metallic structures that are displaced many times during a single day. Besides, the logistic area is located at one of the corners of the factory, while logistic delivery points are located near the different working areas, see Fig. 9.5a. This poses safety risks for robot navigation (collisions with workers or fan cowls should never

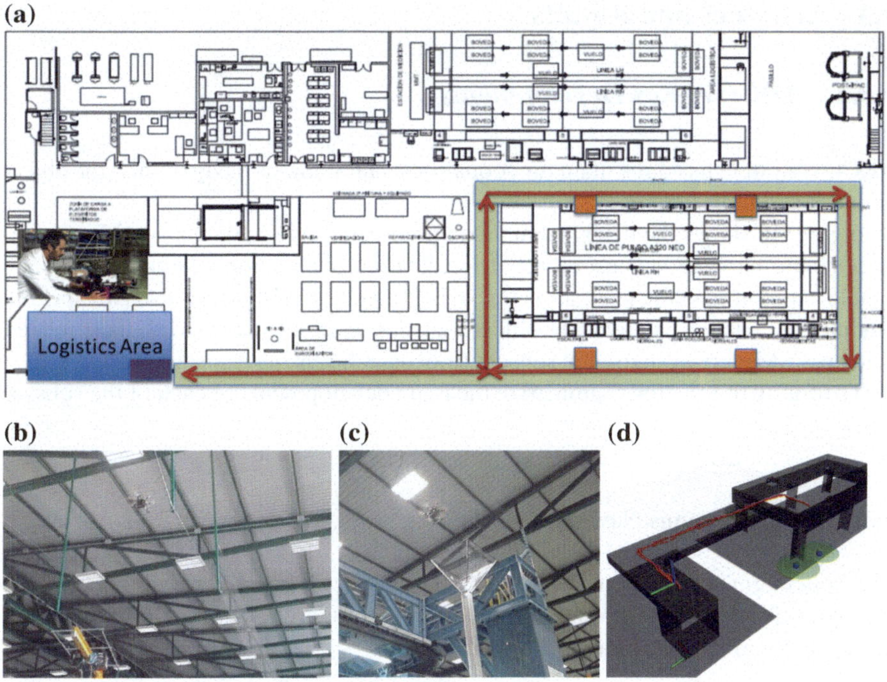

Fig. 9.5 **a** Floor plan of the Airbus D&S CBC fan cowl manufacturing factory. **b** Picture of the aerial robot flying above the safety net. **c** Hopper used for logistic delivery. **d** Map of the area covered by the net to show in real-time the trajectories in the logistic experiments, as well as the robot and tools in the localization experiment

happen, even if the obstacle avoidance algorithms of the aerial robots fail). Aeronautical factories have high density of metallic structures which generate reflections, delays, distortions and multi-propagation effects in radio signals. It is a challenging environment for radio-based tool identification and localization.

To overcome the safety risks, different design approaches were adopted. The aerial robots take off at the logistic area, and always fly above a safety net at an altitude of around 7 m (Fig. 9.5b). Logistic items are provided to workers at the delivery points through a hopper (Fig. 9.5c), in order to minimize risks in human-robot interactions. The behaviour of the system can be visualized in real-time in a 3D map of the area covered by the safety net. For example, Fig. 9.5d shows the trajectory and current position of the aerial robot during a logistic experiment, as well as the tool localization results, with their current associated uncertainties during that same experiment.

9.3 Activities and Results

9.3.1 ARCOW Key Developments

ARCOW system has three main functionalities: autonomous navigation, light-object logistics and tool search.

Figure 9.6 shows the designed architecture for the ARCOW system. Main components are navigation sensors, robot localization, navigation and autopilot. They enable autonomous safe navigation through the scenario. There are also two specific modules for each use case: light-object logistics and tool search, with their related actuators (delivery system) and sensors (tool search sensor) respectively.

The following sections summarize the main developments of each of the subsystems.

9.3.1.1 Autonomous Navigation

This subsystem is the core of the ARCOW system. It enables the possibility of accurate and safe movement through the environment. Figure 9.7 shows the architecture of the autonomous navigation subsystem.

Robot Localization

Robot localization is performed in two phases. Firstly, a sensor-based odometry is performed. It uses the on-board Inertial Measurement Unit (IMU) plus images and point-clouds from an on-board RGB-D camera to estimate the 6 degrees-of-freedom

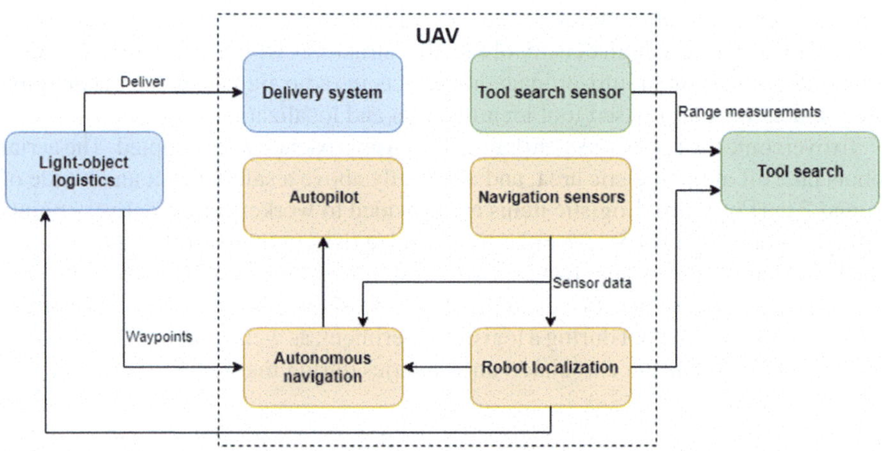

Fig. 9.6 Main modules in ARCOW architecture. The three different functional subsystems are highlighted: autonomous navigation (in orange), light-object logistics (in blue) and tool search (in green)

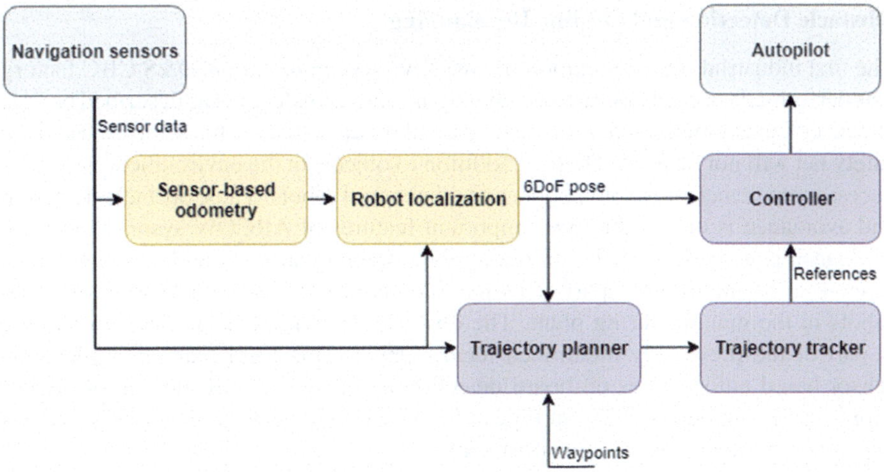

Fig. 9.7 Architecture of the autonomous navigation subsystem. The localization modules are yellow while motion planning modules are purple

pose of the robot. The problem is that this estimation is relative to the initial position of the robot, not maintaining any relationship with a global map of the environment. Thus, a second phase is needed to globally estimate the robot pose. This phase uses Monte Carlo Localization (MCL), which receives the previously estimated visual-inertial odometry together with several raw sensor measurements, which include two RGB-D cameras (one pointing forward and the other pointing backwards) and a radio UWB sensor.

The developed algorithms for robot localization are summarized in Sect. 9.3.2.

Motion Planning

ARCOW motion planning is designed to provide control references based on the robot current pose and a map of the environment. The trajectory planner is actually composed by a global planner and a local planner.

The global planner is able to provide safe paths from robot's location to the desired waypoints, taking into account only a previously built map of the environment, ignoring possible changes in the environment or moving obstacles.

On the other hand, the local planner is continuously updating a local map to enable the detection of new obstacles. It uses the sensor data (i.e. the point-clouds from the cameras) to update the local map by creating, updating or removing these new obstacles. The local planner receives the path calculated by the global planner, and modifies it if necessary.

The last phase of the motion planning is performed by the trajectory tracker and the controller. The first one sends the appropriate references to the latter so that the robot follows the trajectory planned in the previous step. The controller is in charge of translating these references to commands for the autopilot.

More details of the motion planning algorithms are provided in Sect. 9.3.3.

Obstacle Detection and On-line Re-planning

The first industrial implementation of ARCOW system in Airbus D&S CBC factory was performed with only one aerial robot flying above a safety net as described before. However, in next implementations, more aerial robots will fly at the same time and the safety net will not be used. Then, in addition to objects in the environment, it will be necessary to detect and avoid people and other aerial robots. Thus, obstacle detection and avoidance is one of the most important features of ARCOW system. Obstacle avoidance is actually included in the motion planning subsystem, but some details deserve to be mentioned apart. This module enables the safe integration of aerial robots in the manufacturing plant. The obstacle detection and on-line re-planning is part of the previously mentioned local planner. This local planner implements sensor-based autonomous on-board detection and avoidance of moving obstacles. Moreover, it can detect different types of obstacles and perform different strategies depending on the nature of such obstacles:

- Static obstacles. For a new static obstacle (not present in the global map), the strategy is simply to find the shortest path avoiding it with the predefined safety distance from regular obstacles.
- Human workers. In case of detecting a human worker, the robot needs to maintain a higher distance in order to ensure the worker safety. Furthermore, the robot would never fly above the human under any circumstances.
- Other aerial robots. In case of several aerial robots flying in the same environment, they should be aware of each other, and they should be able to avoid each other. The avoidance trajectory is computed with a larger safety distance to consider the motion of the aerial robot that is being avoided. Also, the vertical safety distance was selected with a high value in order to prevent aerodynamic effects and disturbances when an aerial robot flies right above another one.

The obstacle detection is as follows. The navigation sensors, namely the RGB-D cameras, generate point-clouds of the environment. The trajectory planner then compares these point-clouds with the map, and all the points that are not already part of the map, are considered as obstacles. Next, the segmented obstacles are classified into three classes—workers, aerial robots or other obstacles—depending on some descriptors (based on size and shape) and on motion analysis.

Figure 9.8 shows pictures taken in obstacle detection and avoidance tests using a mannequin representing a worker (top left) and another aerial robot (bottom left) as obstacles.

9.3.1.2 Tool Localization

The tool localization is performed integrating the measurements of the distance between the UWB receiver on-board the aerial robot and the tools, each of which was equipped with a UWB tag (see Fig. 9.9).

Fig. 9.8 Obstacles tested in safe navigation experiments: a mannequin (top left) and another aerial robot (bottom left)

Fig. 9.9 UWB receiver mounted on the aerial robot

This functionality can be executed in parallel to the light-object logistic missions since it is computationally inexpensive. Besides, ARCOW system can also perform specific tool search missions, if necessary, for instance at the end of a work shift.

The main challenges for this functionality are the following:

- High density of metallic structures. This is a big issue for the radio-based sensors. Metallic structures cause reflections and occlusions, increasing the measurement noise, creating outliers or blocking the radio signal in some locations. Thus, the sensors and the algorithms need to be robust enough to overcome this challenge.
- Accurate aerial robot localization. Tool localization accuracy highly depends on the precision of the robot location.

The algorithm and some experimental results are summarized in Sect. 9.3.2.2.

9.3.1.3 Light-Object Logistics

This functionality is designed as follows. The aerial robots take off and land from the logistic area in the factory (which is highlighted in Fig. 9.5). When a worker runs out of some material, he/she asks for the small part through a simple HMI running e.g. in a table (a picture of the HMI is shown in Fig. 9.11). This request is received by the logistic operator, who picks the requested materials and places them inside the aerial robot delivery system (see Fig. 9.10). Then, the aerial robot is activated and the transportation mission is commanded. The aerial robot takes off and navigates through the factory to the corresponding hopper, drops the cargo, and comes back to the logistic area.

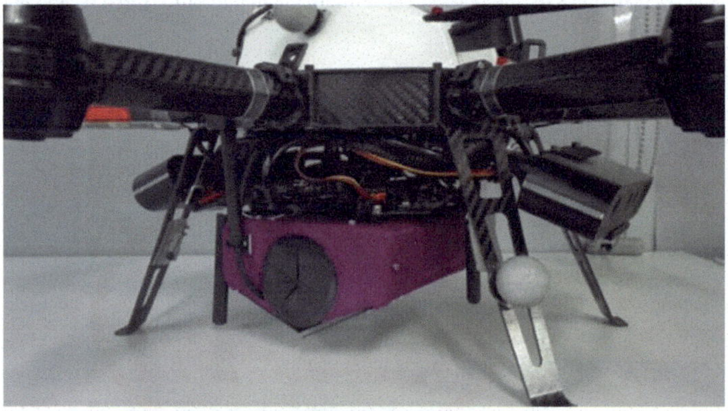

Fig. 9.10 Light-object delivery system mounted on the ARCOW aerial robot

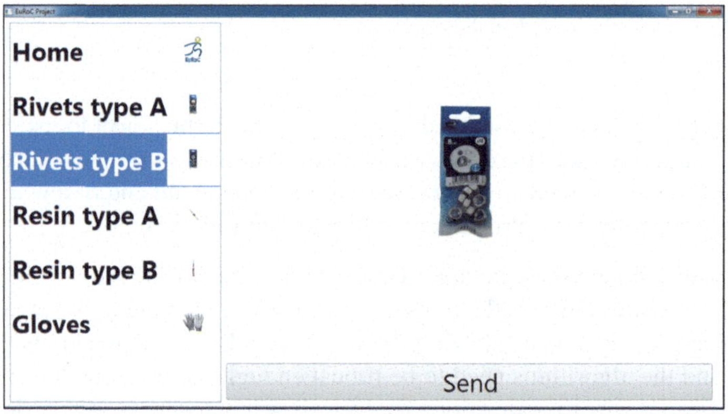

Fig. 9.11 Picture of the HMI developed for requesting and managing the autonomous light-object delivery system

The autonomous delivery of small parts involves the following main challenges:

- Precise point-to-point autonomous aerial robot navigation. It needs the robot to be accurately positioned above the hopper to properly deliver the cargo.
- Global trajectory planning, to find a safe path to the hopper.
- Local reactive obstacle avoidance, to be aware of the changes in the factory and of possible moving obstacles, such as workers or other aerial robots.

9.3.2 *Localization*

The localization problem is essential for building a mobile robotic system, since accurate pose estimation is required for the most basic tasks, even for holding the position. Once this is achieved, the rest of technologies such as navigation and guidance can be implemented in order to determine where the goal positions are and how to reach them.

9.3.2.1 Robot Localization with MCL

Monte Carlo Localization (MCL) is a probabilistic localization algorithm that makes use of a particle filter to estimate the pose of the robot within the map, based on sensor measurements. In general, visual odometry systems are a common option for aerial robot localization because they are lightweight and provide reliable and accurate pose estimations, but only at short term. However, after a period of time, odometry tends to diverge. Instead of using loop closing approaches which can imply an abrupt flight termination, we propose the fusion of visual information with other kind of 3D data acquired from the aerial robot. In our case, UWB sensing is the chosen technology to fuse with a visual odometry system. UWB provides distances to fixed locations that can be used to localize the aerial robot, similar to GPS. Several UWB beacons were installed in the aerial robot surroundings, and another UWB sensor is mounted on-board the robot to gather distance measurements between the different beacons. However, noise typically present in radio-based distance measurements and, mostly, the outliers produced by radio multi-path effect normally make such system not robust enough to use UWB as a stand-alone localization system, especially when high accuracy is required. Moreover, such system does not provide yaw information, hence it is not suitable to constitute a full localization system.

Our system fuses visual odometry together with UWB measurements, making the overall localization system reliable enough at short-term thanks to the visual odometry, and robust enough to cumulative drift thanks to UWB sensing. Furthermore, the algorithms have been implemented having in mind the computational constraints usually found on-board small aerial vehicles, hence they show high efficiency in order to make them suitable for real-time operation. This combination is the main strength of our approach.

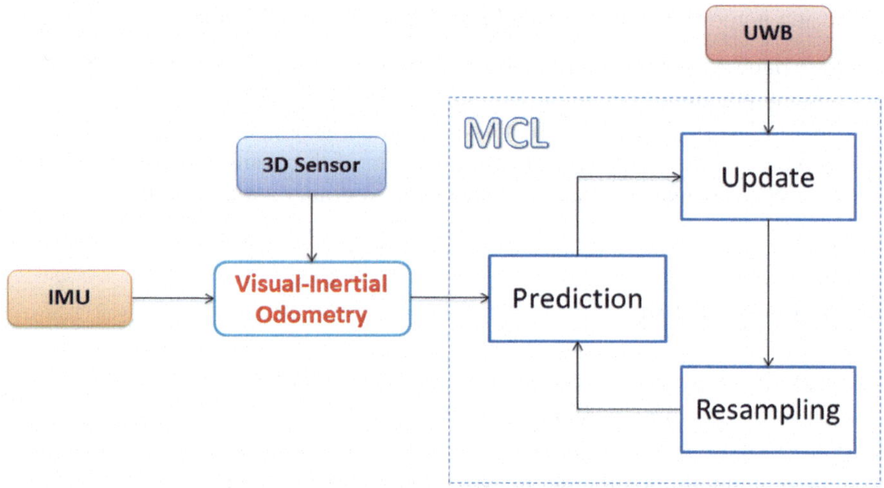

Fig. 9.12 Schematic overview of the localization processing pipeline

The block diagram in Fig. 9.12 depicts an overview of adopted method. The aerial robot localization solution is based on the integration into an MCL of visual odometry coming from an RGB-D camera, 3D point clouds, and distance measurements between the UWB receiver mounted on the aerial robot to UWB tags installed in the scenario at known locations.

The odometry system combines images, 3D point clouds, pose estimations and key-framing in a loosely coupled filter, in order to estimate an accurate and reliable robot localization at short term. As it was previously mentioned, odometry will accumulate errors over time. MCL is able to take care of it by correcting the cumulative drift by two means: comparing the point clouds with a previously known 3D map of the area, and using UWB measurements to fixed beacons in order to readjust localization errors.

The motion model used in MCL is based on the visual odometry computed from RGB-D data developed by the authors [21]. This odometry estimation is applied to the particles of the MCL. The sensor readings for the update step are, on one hand, the point clouds provided by the RGB-D sensor, and on the other hand, the distance measurements to several UWB tags installed in the environment. Then, the weight of each particle is computed according to both types of sensor measurements. The point clouds from the RGB-D camera are transformed to the current pose of each particle in order to find correspondences between such clouds and what the map should look like from that particle's pose. The distance measurements to radio-tags are used to check how well each particle position matches such distances. In order to properly fuse both sensor readings, each particle has a specific weight for each type of sensor, which are later fused into a single weight, as explained in [22].

The particle filter consists of N particles $\mathbf{p}_i$, each one with the following state vector:

$$\mathbf{p}_i = \begin{bmatrix} x \\ y \\ z \\ \psi \end{bmatrix}, \tag{9.1}$$

where ψ refers to the yaw angle of the aerial robot. The roll and pitch angles of the robot are not included into the state definition, since we can assume that such angles are available and accurate enough in aerial robots through the use of its on-board IMU. This simplifies the computational complexity of the algorithms, making our approach suitable for real-time on-board computation.

Experimental Validation

The MCL-based localization was extensively tested at CATEC indoor testbed, especially for the purpose of showcasing the use cases during Stage 2.b. This subsection provides experimental results from the estimated localization compared to ground truth data obtained from the testbed motion capture system.

The following elements constitute the full experimental setup:

- Aerial robot with two on-board RGB-D cameras (Orbbec Astra), an IMU and a UWB sensor.
- 3D map of the environment obtained through one of the RGB-D cameras in a previous flight, using the motion capture system as localization data for projecting the 3D point-clouds.
- Three UWB sensors installed in the indoor testbed at known positions (see Fig. 9.13).

The main objectives of the experiments are twofold; demonstrate that the localization accuracy is enough to safely navigate through the environment, and demonstrate the suitability of the algorithms for real-time computation. The estimated aerial robot position and yaw during one the flights can be seen in Fig. 9.14, both from the visual odometry (blue) and the Monte Carlo Localization (green), as well as the ground-truth data provided by the testbed's motion capture system (red). Roll and pitch angles are not included in the plots because, as previously stated, they are observable from the on-board IMU. It can be seen how the estimations from the visual odometry slowly accumulate errors and drifts, while the estimations from our MCL closely follow the ground-truth position and yaw during the whole flight. It is also worth to mention that this estimation does not drift with time and the errors are approximately bounded.

Table 9.1 shows some of the results obtained in these experiments. The RMS errors are approximately 0.4m in x and y axes, while the error in z is lower than 0.2m. The RMS error in *yaw* angle is around 10 degrees. Errors in x and y are slightly higher than in height due to the large distances traveled through these axes during this experiment. Nevertheless, the overall RMS error in position is 0.32m, which can be considered acceptable for robot navigation.

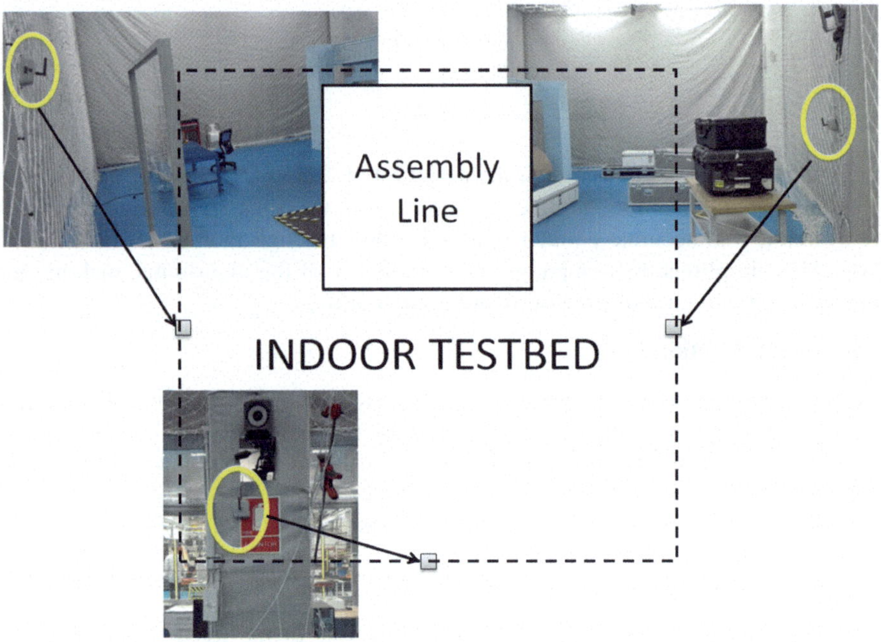

Fig. 9.13 CATEC indoor testbed with approximate location of UWB beacons

Table 9.1 RMS localization errors

	x (m)	y (m)	z (m)	yaw (rad)
Odometry	2.47	2.53	0.19	0.88
MCL	0.36	0.39	0.17	0.18

9.3.2.2 Tool Localization

Using the pose obtained in Sect. 9.3.2.1 and the radio range measurements gathered from the different tools, the aerial robot is capable of detecting and locating the missing tools in the scenario. The proposed method has two sequential stages that are executed for each missing tool:

- First, a Particle Filter (PF) is used to obtain an approximate initial location of the tool. This is necessary since the range-only measurements inherently give multi-modal solutions for beacon localization until convergence is found [23].
- Second, an Extended Kalman Filter (EKF) is used to refine the tool location estimation. This EKF is much more efficient to evaluate than the PF, and thus improves the tool localization with lower computational cost than the PF.

The proposed method employs one PF-EKF combination for each missing tool. Different tools have different radio-tag IDs, thus solving the measurement association

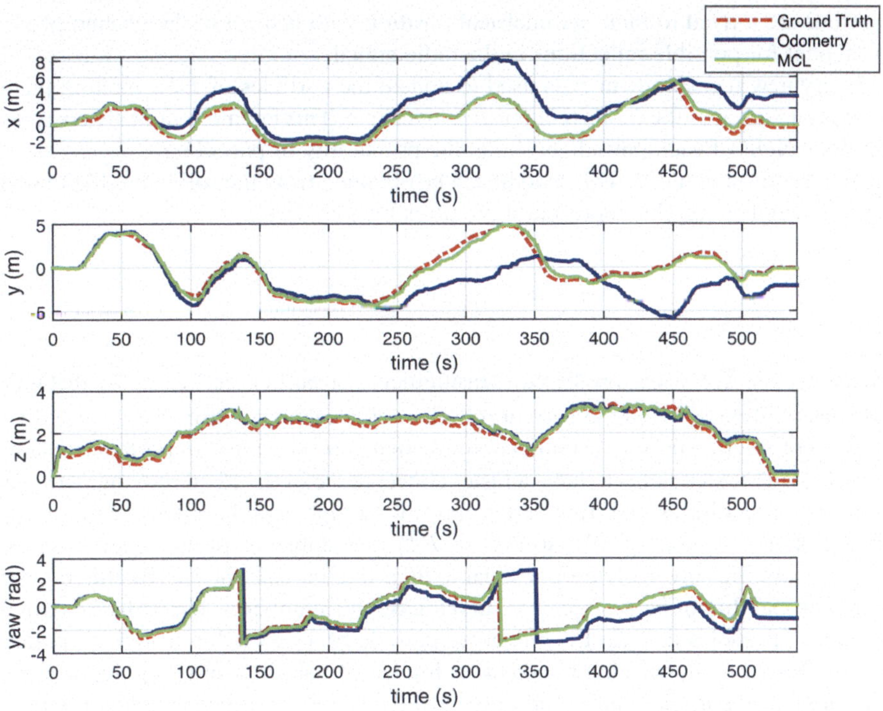

Fig. 9.14 Localization results in one experiment in the CATEC testbed. The aerial robot ground-truth data captured by the motion capture system is shown in red color. The estimated robot localization using only the visual odometry approach is in blue. The estimated robot localization using the full MCL approach is shown in green

problem. Given that individual PFs have faster convergence than more complex multi-tool PFs, this also reduces significantly the involved computational burden. Consequently, the tool search can be performed by the aerial robot while also doing other tasks (such as logistics), running simultaneously with other programs on-board the aerial robot.

The PF Stage

Let z_i be the received range measurement from tool i. Assume that the current position of the UBW receiver onboard the aerial robot, X_r, is known. X_r can be obtained as in Sect. 9.3.2.1. When the first measurement z_i is received, PF_i -the Particle Filter for radio tag i- is initialized. Each particle $\chi_i^{[j]}$ in PF_i represents an hypothesis of the 3D position of the tool i, X_i. The initialization is done by drawing the particles from a 3D annular distribution with mean z_i and width 3σ, with σ the standard deviation of the range measurements noise. Additionally, particles located higher than 3 meters are rejected (since the work area is below that altitude), and particles located beneath the floor are moved to their symmetrical position with respect to the ground plane, to account for possible reflections in the radio signal.

Every new measurement z_i is used to update the particles of PF_i, making them condensate towards the tool's real position overtime. This update is done by computing the weight of each particle $\omega_i^{[j]}$, i.e. the probability of particle $\chi_i^{[j]}$ to represent the real position of tool i. This weight can be obtained from the following likelihood function (higher weights represent more probable particles):

$$\omega_i^{[j]} = \frac{1}{\sigma\sqrt{2\pi}} e^{\frac{(d(\chi_i^{[j]}, X_r) - z_i)^2}{2\sigma^2}}, \tag{9.2}$$

where $d(\chi_i^{[j]}, X_r)$ is the predicted measurement for particle $\chi_i^{[j]}$, i.e. the distance between particle $\chi_i^{[j]}$ and the current robot's radio beacon position X_r.

Each of the PFs in our system also implements the so-called importance resampling [24]. This step helps to have a faster convergence of the PF, removing particles with low probabilities and duplicating those with higher probabilities. After every update of the weights of all particles in PF_i, a number of particles are rejected with probabilities inversely proportional to their likelihoods (the less likelihood, the higher probability of being rejected). In our case this number is M=10% of the total amount of particles in PF_i. Each rejected particle becomes a duplicated of another one (in this case, higher likelihoods mean higher probabilities of being duplicated). The duplicated particle is also randomly moved by a few centimeters to favor finding other possible solutions.

Finally, when the maximum eigenvalue of the covariance matrix for tag i (3D covariance of the position and weights of all particles in PF_i) is lower than a certain threshold, PF_i is considered as converged to a Gaussian unimodal distribution. When PF_i has converged to an unimodal distribution, the tool localization X_i is estimated as the weighted mean of all particles:

$$X_i = \sum_{j=1}^{M} \omega_i^{[j]} \chi_i^{[j]} \tag{9.3}$$

Once X_i is estimated, it is used in combination with the covariance matrix of the particles to initialize an EKF that will henceforth refine the tool's estimated position. This PF-EKF structure reduces drastically the computational burden of the system, but still allows to improve the tool's localization estimation with the subsequent measurements.

EKF-Based Refinement

After PF convergence for tool i, a new EKF, EKF_i, is initialized as explained in Sect. 9.3.2.2. This EKF assumes that the missing tool is static. Thus, its prediction model will be $X_{i,t} = X_{i,t-1}$, where $X_{i,t}$ is the predicted position of tool i at time t.

The observation model is the expected distance measurement between the tag i and the robot's radio beacon. This is equal to the Euclidean distance between the aerial robot and the missing tool:

$$z_i = |x_i - x_r| \tag{9.4}$$

Given the prediction and observation model, the implementation of the EKF is straightforward. More complex models can be used to improve convergence by integrating inter-beacon measurements (distances between different tools) such as in [23], at the cost of gathering them, but the proposed PF-EKF solution has proven accurate enough for the required use case.

Experimental Validation

The tool localization system was implemented on-board the aerial platform and validated in two different scenarios: CATEC indoor testbed and the Airbus D&S CBC factory. Both scenarios gave similar radio distance measurement noise, but the CBC factory provided a significantly higher amount of outlying measurements. This is due to the presence of more obstacles, many of them metallic, that affect the radio measurements as explained in Sect. 9.2.3.3. To overcome this issue, a median-based outliers filter (explained in [25]) was implemented, satisfactorily eliminating almost 95% of the outliers without affecting inlying measurements.

An example of a tool localization experiment is found in Fig. 9.15. It shows several steps in the execution of the PF-EKF method for tool search and localization including: (a) particle initialization; (b) PF convergence; (c) EKF initialization and; (d) resulting tool position estimation. The aerial robot take-off and landing location is shown at the bottom-right corner of the images. The path is represented by red points. Blue circles represent the ground-truth position of the tools. The particles of each PF are deployed when the first measurement from that radio tag is received, Fig. 9.15a. After some measurements have been received, the corresponding PFs converge and are represented with an ellipse along with the particles. If convergence is below a given threshold (the maximum eigenvector is <0.5 m), the subsequent EKF is initialized, and is represented only with an ellipse. When the robot lands, the tools' estimated locations are provided to the workers in a geometrical 3D representation, see Fig. 9.15-d. Particle aggregations far from the ellipses represent lowly weighted particles that will be resampled in future steps.

The performance of the tool localization system was analyzed in both the CATEC and CBC scenarios. The deployment of the tools was done in many different configurations and places. Only radio tags hidden below fan cowls were more difficult to find than other configurations due to the high ratio of outliers. However, this is an unlikely situation in the factory use case. Table 9.2 shows the mean localization error and convergence time of all the tools, both in the CATEC and CBC scenarios, and also with and without the outliers filter. Convergence time is interpreted as the time spent between the integration of the first radio measurement of tool i, and the initialization of EKF_i (which requires previous convergence in PF_i).

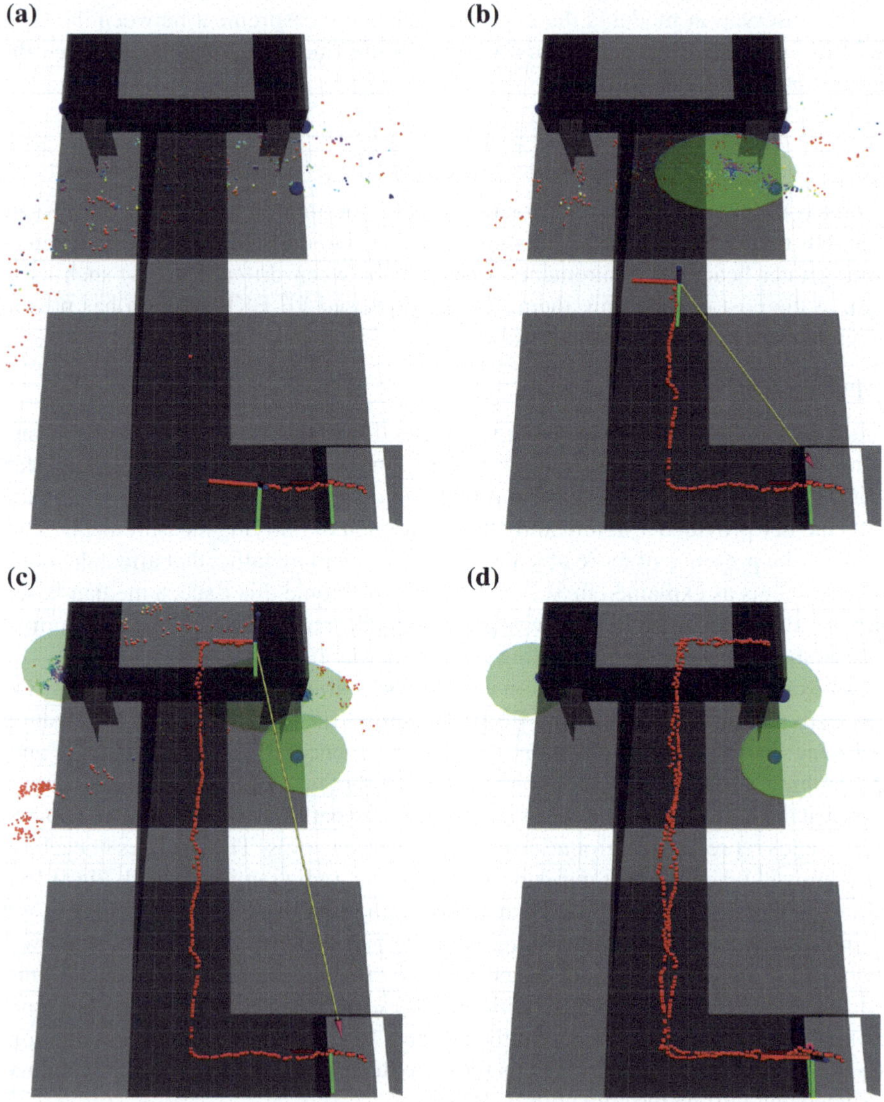

Fig. 9.15 Several steps in the execution of the PF-EKF method for the search and localization of 3 tools: **a** PF particle initialization; **b** PF convergence; **c** start of the EKF stage one of the tools, and; **d** EKF final location estimation for all tools

Table 9.2 Average results in the experiments

Scenario	CATEC scenario		CBC scenario	
Outlier filter	No	Yes	No	Yes
Accuracy (m)	0.31	0.28	2.15	0.34
Conv. Time (s)	21	21	130	41

It can be seen in Table 9.2 that the results in the CATEC scenario have little or no difference whether the outlier filter is used or not. However, the CBC scenario does present important differences: if the outlier filter is not used, both the accuracy and the convergence time of the method suffer significantly. This is due to the presence of obstacles and metallic structures in the CBC factory, which results in a high number of outliers. On the other hand, even when using the outlier rejection filter, the convergence time in the CBC scenario is almost the double than in the CATEC scenario. This is due to the larger size of the CBC factory when compared to the CATEC scenario, and also to the lower operational speed of the aerial robot in the CBC factory, which was decreased to accommodate safety requirements. However, the tool search system would still reduce the current tool search operational time by a $\approx 70\%$, thus validating the proposed method in the final scenario.

9.3.3 *Motion Planning*

This section describes the algorithms implemented in the trajectory planner and tracker modules presented in Sect. 9.3.1. The trajectory planner consists of two different planning algorithms: the global planner and the local planner. Both make use of an enhanced version of the Lazy Theta* algorithm, which we called Weighted Lazy Theta*. It is able to maximize the probability of success while reducing the planning time and always obtaining the same re-planned trajectory from an initial target position for each map.

9.3.3.1 Weighted Lazy Theta*

The main disadvantage of A* algorithms is that paths are formed by grid edges. Theta* is based on A*, without paths constrained by graph edges. Each expanded vertex performs a line-of-sight check for each expanded visible neighbor that allows the parent of this vertex to be any other graph vertex, and thus it finds realistic and short looking paths. Lazy Theta* reduces the number of line-of-sight checks in comparison to the Theta*, finding paths faster but without increasing the length of the path.

The Lazy Theta* algorithm has two main advantages in our case. First, it reduces the planning time. Second, it can solve restrictions imposed by the physical location of sensors. We implemented the Weighted Lazy Theta* algorithm, a variation in which the distance is heuristically weighted as follows:

$$h(s) = w * c(s, s_{goal}), \tag{9.5}$$

where the h-value is the straight line distance from a vertex s to the goal vertex s_{goal}, which is weighted with positive values of weights w. If the value of w decreases, the search expands to more vertices, so the algorithm would be able to provide shorter paths at the expense of increasing the planning time. Otherwise, if the value of w increases the affected number of vertices is reduced, greatly decreasing the run time and without implying a significant increase in the length of the paths.

In our implementation the cost function $c(s, s_{goal})$ is taken as the straight line distance but weighting the z coordinate with z_w as follows:

$$c(s, s_{goal}) = \sqrt{(s_{goal,x} - s_x)^2 + (s_{goal,y} - s_y)^2 + z_w(s_{goal,z} - s_z)^2}, \tag{9.6}$$

where $s_{goal} = [s_{goal,x}, s_{goal,y}, s_{goal,z}]$ is the goal position, $[s_x, s_y, s_z]$ is the current robot location and z_w is the weight to differently consider the z coordinate. Increasing the height weight z_w reduces the Z-axis exploration and the horizontal planning is prioritized decreasing the algorithm cost.

The complete motion planning system makes use of two occupancy maps. The global map, which is used by the global planner, and is built from an octree with known obstacles and map. The local map is used by the local planner and is continuously updated from the point cloud. This point cloud should be pre-processed to remove outliers and reduce the number of points. While the global map is static, the local map is updated with an auxiliary map created with the pre-processed point clouds.

Both maps are voxel-based matrices obtained inflating the occupied points from the octree or from the point cloud. Inflaction is performed vertically and horizontally by half the robot size, plus a safety factor.

9.3.3.2 Global Planning

The global planner consists of a Weighted Lazy Theta* algorithm that computes the path from the current robot pose to the goal pose in the global map. This path is then post-processed to obtain a global trajectory from the current robot pose to the final goal pose.

The global trajectory is calculated from the computed path based on the following parameters: the chosen angular and linear main velocities (yaw), the maximum distance between waypoints and a desired attitude reference evolution.

In global trajectory planning two yaw tracking methods have been used to modify the yaw between waypoints. The first is the simplest option: before the aerial robot starts moving it changes its yaw to the yaw reference of the following waypoint in the segment. This method is not the safest solution because it may generate blind movement segments. In the second method before the aerial robot starts moving it changes its yaw so that the robot X axis is oriented to the following waypoint. When the robot reaches the following waypoint, it changes its yaw using the yaw reference of that waypoint. This combination of yaw tracking methods has two main objectives:

- To confirm that the trajectory waypoints are close enough to each other so that time references obtained from main velocities and attitude references are realistic.
- To have the origin and the final goals of a potential re-planning as close as possible to each other.

The global planner includes a state machine over the local planner and the trajectory tracker in order to ensure that the goal is reached or to cancel the trajectory execution if the re-planning fails a given number of times.

9.3.3.3 Local Planning

The local planner is a two-step algorithm that handles the global trajectory—obtained from the global planner– according to the information in the local occupancy map.

In the first step it updates the local map and then checks if re-planning is really necessary (i.e. if the presence of obstacles across the trajectory is detected in the local plan). It checks if all the waypoints from the current robot pose to the last waypoint inside the last environment map have line of sight with his predecessor. Lacks of line of sight between consecutive waypoints are interpreted as obstacles in the trajectory. If one or more than one wayspoints do not have line of sight with his predecessor, re-planning is necessary.

The second step is re-planning. The objective is to avoid the new obstacles by getting a new and safe global trajectory. It makes use of Weighted Lazy-Theta* to obtain a collision free subtrajectory which replaces the trajectory segment where the obstacle has been detected and thus obtains a new and safe global trajectory free of obstacles. Figure 9.16 shows data from two experiments where re-planning has been performed using the information from the local map (in yellow/blue) color. The global map is in green.

9.3.3.4 Trajectory Tracking

The objective of the trajectory tracking module is to ensure that all the trajectory waypoints are safely reached at their specific times. In order to ensure that we can reach the waypoints in all the axes with enough time, the time references should be

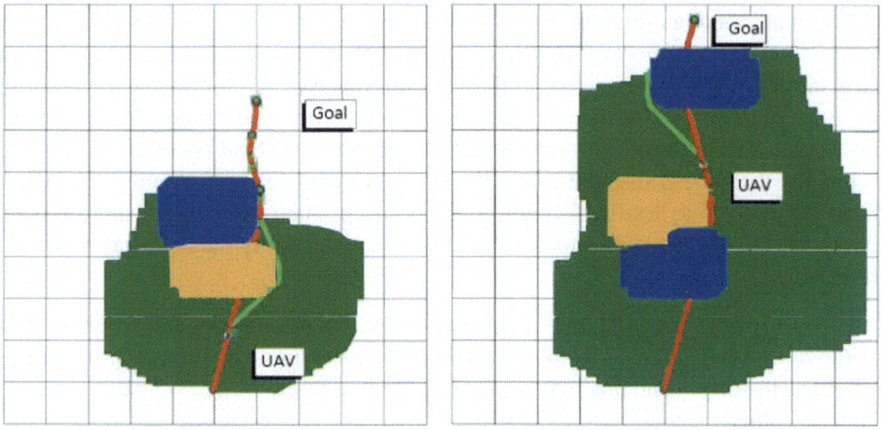

Fig. 9.16 Consecutive re-planning subtrajectories (green line and green cubes) of the local planner along a global trajectory (red line and orange cubes) given two waypoints, one around the middle of the grid and another one at the top. The global and local maps are respectively shown in green and yellow/blue, respectively. The left image shows the aerial robot avoiding a known wall that is placed incorrectly in the global map

the highest value among the minimum necessary time to achieve the required linear and angular movement in each axis (X, Y, Z, Yaw).

Also, the progression between waypoints should be smooth and without overhead. The continuous trajectory can be generated by means of interpolation, but that does not guarantee to comply with the safety distances to the obstacles. Hence, it is necessary to have a module—the trajectory generator—that transforms discrete trajectory waypoints into a continuous trajectory. The trajectory generator computes velocity commands for each robot axis along each trajectory segment according to the velocity at the beginning of the line segment and based on a linear model of the aerial robot. The velocity generator commands constant speed in each segment at each axis (X, Y, Z, Yaw).

All trajectory segments in the trajectory are treated similarly and the change of velocity references is anticipated to each previous trajectory segment.

9.4 Conclusion

There is a clear need in many factories to increase their level of automation. The *Industry 4.0* paradigm intends to massively introduce information and automation technologies in industrial processes. Up to now, the greater majority of the robot co-workers in industrial sectors that have been proposed refer mainly to manipulators, ground robots or even exoskeletons. In this chapter we describe our participation in EuRoC with our project ARCOW, the Aerial Robot Co-Worker in Plant Servicing, in which we have explored the development, implementation and validation of aerial robots as co-workers in manufacturing tasks in the aeronautic industry. University of Seville, FADA-CATEC and Airbus D&S integrated the consortium of ARCOW project, which was coordinated by the University of Seville.

Aerial robots provide interesting advantages versus traditional ground robots in manufacturing processes. First, the floor of factories are often very dense and dynamic which constrain the motion of ground robots, whereas the upper part in factories is rather sparse and static, more suitable for aerial robot navigation. Besides, safety constraints impose low-speed motions for ground robots, reducing operational efficiency, whereas this constraint is not present for aerial robots. It should be taken into account that aerial robot regulations do not apply in indoors. On the other hand, introducing aerial robots in manufacturing processes interesting challenges that should be addressed. The system should include GNSS-denied precise long-term robot pose estimation functionalities as well as efficient planning methods suitable for tight and dynamic environments. From an operational and implementation perspective, the system should ensure safe aerial robot navigation in tight and crowded environments.

ARCOW was focused on two use cases, selected due to their importance to the aeronautical manufacturing sector: (1) use of aerial robots to improve operational efficiency and reduce costs in logistic operations of light objects, and (2) use of aerial robots to improve operational efficiency in detection and localization of missing tools.

The chapter described the methodology we adopted in ARCOW as well as the designed architecture, main methods and their validation in different scenarios in the subsequent stages of EuRoC.

The validation pointed out the interest for the proposed application in industrial plants. The first industrial application described in this chapter, by using nets to ensure the safety, will lead to other implementations without the net, by using the collision detection and avoidance functionality, also described in this Chapter, and eventually other platforms with intrinsic structural safety.

Field testing is one of the most expensive tasks when developing an autonomous system because the aerial platform must be fully operational before carrying out flight experiments, a space for testing must be accessible, and specific qualified staff must be available (including a safety pilot). Besides, if testing outdoors, the weather conditions can be a constraint and, depending on national regulations, a valid license for the aerial robot pilot might be required. Fortunately, the latter requirement did not play a major role in our case, since all the experiments took place in indoor environments (Fig. 9.17).

An important factor to take into consideration when testing aerial robots is their time of flight. Degradation of batteries is a significant issue that we encountered during extensive test campaigns, leaving us with limited flight times. Moreover, the aerial platform under test was a research prototype from *Ascending Technologies*, which made us depend heavily on the manufacturer, whose late response forced us to manage custom battery replacements.

Fig. 9.17 CATEC indoor testbed, where most of the testing and validation has taken place

Simulation environments such as Gazebo[1] have been very useful for testing the behavior and integration of different modules, especially those related to motion planning. However, simulations are still limited in order to provide accurate information regarding real world scenarios or behaviors of actual systems.

The *node_manager* developed by FKIE[2] has been a very convenient tool that was extensively used during almost all the testing experiments. It is a graphical interface to manage nodes, topics, services, parameters and launch files in a ROS network. It seamlessly handles the discovery of running ROS masters in the local network and the synchronization of the local ROS master to the discovered remote masters. This is very useful because all the algorithms are running in the on-board computer, and, at the same time, the ground computer can run visualization and debugging tools in order to watch live sensor data, or check the expected behavior of specific algorithms that are under test.

The ability to evaluate the performance of the aerial robot in an experiment is a key factor in field testing. Sometimes a tight time schedule has forced us to carry out extensive test campaigns, during which it was not unusual to adjust several parameters *on the fly*. This leaves room for subjectivity, since human errors can easily arise. It should be a good idea to develop automated methods to quantitatively assess recorded data in order to evaluate the aerial robot performance.

[1] http://gazebosim.org.

[2] http://wiki.ros.org/node_manager_fkie.

References

1. Haddadin, S., Suppa, M., Fuchs, S., Bodenmüller, T., Albu-Schäffer, A., Hirzinger, G.: Towards the robotic co-worker. In: Cédric Pradalier, Siegwart, R., Gerhard, H., (Eds.) Robotics Research, pp. 261–282. Springer Berlin Heidelberg (2011)
2. Sauppé, A., Mutlu, B.: The social impact of a robot co-worker in industrial settings. In: Proceedings of 33rd ACM Conference on Human Factors in Computing Systems, CHI '15, pp. 3613–3622 (2015)
3. Davison, A.J., Reid, I.D., Molton, N.D., Stasse, O.: Monoslam: real-time single camera SLAM. IEEE Trans. Pattern Anal. Mach. Intell. **29** (2007)
4. Achtelik, M.W., Lynen, S., Weiss, S., Kneip, L., Chli, M., Siegwart, R.: Visual-Inertial SLAM for a small helicopter in large outdoor environments. In: *IEEE/RSJ IROS2012*, vol. 10, pp. 2651–2652 (2012)
5. Weiss, S., Scaramuzza, D., Siegwart, R.: Monocular-SLAM-based navigation for autonomous micro helicopters in GPS-denied environments. J. Field Robot. **28**:854–874 (2011)
6. Honegger, D., Meier, L., Tanskanen, P., Pollefeys, M.: An open source and open hardware embedded metric optical flow CMOS camera for indoor and outdoor applications. In: IEEE ICRA2013, pp. 1736–1741 (2013)
7. Mebarki, R., Lippiello, V.: Image moments-based velocity estimation of UAVs in GPS denied environments. IEEE International Symposium on Safety, Security, and Rescue Robotics, pp. 1–6 (2014)
8. Kneip, L., Chli, M., Siegwart, R.Y.: Robust real-time visual odometry with a single camera and an IMU. In: Proceedings of the British Machine Vision Conference. BMVA Press (2011)
9. Geiger, A., Ziegler, J., Stiller, C.: Stereoscan: Dense 3d reconstruction in real-time. In: IEEE Intelligent Vehicles Symposium, pp. 963–968 (2011)
10. Endres, F., Hess, J., Engelhard, N., Sturm, J., Cremers, D., Burgard, W.: An evaluation of the RGB-D SLAM system. In: IEEE ICRA2012, pp. 1691–1696 (2012)
11. Kerl, C., Sturm, J., Cremers, D.: Robust odometry estimation for RGB-D cameras. In: IEEE ICRA2013, pp. 3748–3754 (2013)
12. Pomerleau, F., Colas, F., Siegwart, R., Magnenat, S.: Comparing ICP variants on real-world data sets. Autonom. Robots, 133–148 (2013)
13. Nuchter, A., Lingemann, K., Hertzberg, J.; Cached k-d tree search for ICP algorithms. In: Sixth International Conference on 3-D Digital Imaging and Modeling (3DIM 2007), pp. 419–426 (2007)
14. Marden, S., Guivant, J.: Improving the performance of ICP for real-time applications using an approximate nearest neighbor search. In: Australasian Conference on Robotics and Automation (2012)
15. Kümmerle, Rainer, Ruhnke, Michael, Steder, Bastian, Stachniss, Cyrill, Burgard, Wolfram: Autonomous robot navigation in highly populated pedestrian zones. J. Field Robot. **32**, 09 (2014)
16. Lavalle, S.M.: Rapidly-exploring random trees: a new tool for path planning. Technical Report Computer Science Department, Iowa State University (TR 98–11), 05 1999
17. Jaillet, I., Juan, C., Siméon, T.: Transition-based RRT for path planning in continuous cost spaces. In: IEEE/RSJ IROS2008, pp. 2145–2150, 09 (2008)
18. Torres-Gonzalez, A., Martinez-De Dios, J.R., Ollero, A.: Efficient robot-sensor network distributed SEIF range-only SLAM. IEEE ICRA2014, pp. 1319–1326 (2014)
19. Torres-González, A., Dios, Jose Ramiro Martinez-de, Ollero, A.: Range-only SLAM for robot-sensor network cooperation. Autonom. Robots **42**(3), 649–663 (2018)
20. Burri, M., Nikolic, J., Gohl, P., Schneider, T., Rehder, J., Omari, S., Achtelik, M.W., Siegwart, R.: The EuRoC micro aerial vehicle datasets. Int. J. Robotics Res. **35**(10), 1157–1163 (2016)
21. Perez-Grau, F.J., Ragel, R., Caballero, F., Viguria, A., Ollero, A.: Semi-autonomous teleoperation of UAVs in search and rescue scenarios. In: 2017 International Conference on Unmanned Aircraft Systems (ICUAS), pp. 1066–1074 (2017)

22. Perez-Grau, F.J., Caballero, F., Viguria, A., Ollero, A.: Multi-sensor three-dimensional monte carlo localization for long-term aerial robot navigation. Int. J. Advanc. Robotic Syst., **14**(5) (2017)
23. Torres-González, A., Martinez-de Dios, J.R., Ollero, A.: Range-only SLAM for robot-sensor network cooperation. Autonom. Robots, pp. 1–15 (2017)
24. Thrun, Sebastian, Fox, Dieter, Burgard, Wolfram, Dellaert, Frank: Robust monte carlo localization for mobile robots. Artif. Intell. **128**(1–2), 99–141 (2001)
25. Paneque, J.I., Torres-González, A., Ramiro Martínez-de Dios, J., Ramón Astorga Ramírez, J., Ollero, A.: Autonomous localization of missing items with aerial robots in an aircraft factory. In: Iberian Robotics Conference, pp. 179–189. Springer, Cham (2017)

The manufacturer's authorised representative in the EU is Springer Nature Customer Service Centre GmbH, Europaplatz 3, 69115 Heidelberg, Germany. If you have any concerns regarding our products, please contact ProductSafety@springernature.com

Printed and bound by CPI Group (UK) Ltd, Croydon, CR0 4YY
22/09/2026
02221057-0004